MAPS AND MAP-MAKERS

1   TITLE-PAGE TO R. DE HOOGHE'S *ATLAS MARITIME*, 1693

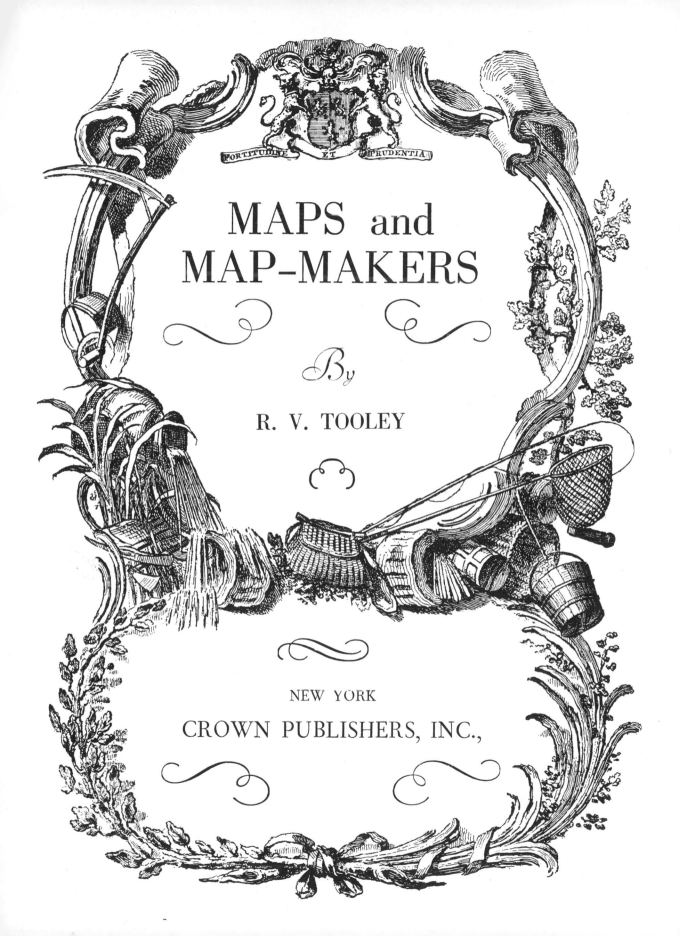

MAPS and MAP-MAKERS

By

R. V. TOOLEY

NEW YORK

CROWN PUBLISHERS, INC.,

PORTITUDINE ET PRUDENTIA

To My Sister

MARIAN JOYCE TOOLEY

© R. V. Tooley

First published in U.S.A. 1978
by Crown Publishers, Inc.

Library of Congress Cataloging in Publication Data

Tooley, Ronald Vere, 1898–
Maps and map-makers.

Includes bibliographies and index.
1. Cartography–History. I. Title.
GA201.T6 1978   526.8'09   77–16227
ISBN 0–517–53303–0

Preface

A CONSIDERABLE literature has grown up in recent years on the subject of Cartography, but by far the greater part of this output is highly specialised, and for the most part scattered in the publications of various learned societies not often obtainable by the student or collector, or in limited and expensive monographs. There has so far been no general work dealing with the whole subject. It is to fill this desideratum that this book has been compiled, the aim being to give information and illustrations of the principal map-makers and map publishers and their work from the earliest times to the nineteenth century, combining an appreciation of the popular decorative side of early maps with historical and bibliographical notes.

No one section has been treated exhaustively, as it is intended only as a preliminary guide to students and collectors. The book is arranged in schools of geography, the Classics, Italian, Dutch, French, and English, with brief notes on the principal map-makers in each group and a list of their main productions. It embraces the whole world, and for those who wish to pursue a more detailed study of a particular area or period, the works of the various specialists have been cited.

I wish to express my recognition to all previous writers on the subject, particular reference to whose work has been given at the end of each chapter, and my thanks to Dr. Lynam and the staff at the British Museum, the Librarians of the Royal Geographical Society and Greenwich Museum for their courtesy and help at all times, and to Messrs. Francis Edwards, The Campden Gallery, Hodgson, Maggs Bros., Sotheby, and Henry Stevens, Son and Stiles for the facilities they have given me to examine maps in their possession; to my sister for corrections in Chapter II, and to Miss Joyce Simmonds for her care in typing and overseeing the manuscript.

LONDON, 1949   R. V. T.

NOTE TO THE SIXTH EDITION

For this, the sixth edition of what has now become a standard work, a fresh general introduction has been provided, and the lists of Authorities at the ends of chapters have been brought up to date.

LONDON, 1978   The Publishers

List of Illustrations

Acknowledgment

The Author and Publishers would like to acknowledge their great indebtedness to Messrs. Francis Edwards Ltd., of Marylebone High Street, the Misses Kelly of The Campden Gallery, of Kensington Church Street, and Messrs. Maggs Brothers Ltd., of Berkeley Square, for generously lending the original maps from which the majority of the illustrations were made.

So far as the remainder of the illustrations are concerned, Figs. 8, 15, 18, 20, 22, 42, 43, 68, 86, 97 and 103 are from originals in the Map Room at the British Museum, and are reproduced by permission of the Trustees; and Fig. 9 from a photograph belonging to the Royal Geographical Society, is reproduced, by permission of the authorities, from the Mappa Mundi at Hereford Cathedral.

The colour plates are all taken from contemporary colour maps; those on Figs. 36 and 47 from a copy of John Speed's *Theatre of the Empire of Great Britain* now in the possession of Dr. Eric Gardner.

The design on the title-page comes from a cartouche on Eman. Bowen's *Shrop Shire* (1760).

2 *The instruments of the surveyor*

Contents

General Works of Reference

BAGROW (Leo). *Ortelli Catalogus Cartographorum*. Gotha, Perthes, 1928.
—— *Geschichte der Kartographie*. Berlin, Safari Verlag, 1951.
—— English edition, *History of Cartography*, revised and enlarged by R. A. Skelton. London C. & A. Watts, 1964. Reprinted by Spring Books, 1965.
BONACKER (W.). *Kartenmacher aller Länder und Zeiten*. Stuttgart, Hiersemann, 1966.
BRITISH MUSEUM. *Catalogue of Printed Maps, Charts and Plans*. Photolithographic edition to 1964. London, Trustees of the British Museum, 1967. 15 vols.
BROWN (Lloyd A.). *The Story of Maps*. Boston, Little Brown & Co., 1949.
CRONE (G. F.). *Maps and their Makers*. London, Hutchinson, 1966.
FITE (E. D.) & FREEMAN (A.). *A Book of Old Maps*. Mass., Harvard U.P., 1936.
FORDHAM (Sir H. G.). *Maps, their History, Characteristics and Uses*. Cambridge U.P., 1927.
HARMS (Hans). *Künstler des Kartenbildes*, portraits and short biographies of 102 map-makers. Oldenburg, Völker, 1962.
HUMPHREYS (A. L.). *Old Decorative Maps and Charts*. London, Halton & Truscott Smith, 1926. With catalogue of the Macpherson Collection.
—— Revised edition by R. A. Skelton. Staples Press, London, 1952. Without the catalogue of the Macpherson Collection.
Imago Mundi. A review of early cartography founded by Leo Bagrow. Amsterdam, N. Israel. 20 vols., 1935-69. Issued roughly one vol. per year. Continuing.
O. KOEMAN (Dr. I. C.). *Collections of Maps and Atlases in the Netherlands*. Leyden, 1961.
—— *Atlantes Neerlandici*. Vols. I and II, Amsterdam, T.O.T., 1967-9.
LEITHAUSER (J. G.). *Mappae Mundi*. Berlin, Safari Verlag, 1958.
LISTER (Raymond). *How to Identify Old Maps and Globes*. London, Bell, 1965.
LYNAM (Edward). *The Mapmaker's Art*. London, Batchworth Press, 1953.
LIBRARY OF CONGRESS. *List of Geographical Atlases*. Washington, Lee Phillips, 4 vols., 1909–20. Continuation Le Gear, Vols. 5 and 6, 1958–63.
MAP COLLECTORS' CIRCLE. Edited by R. V. Tooley. 11 vols., 2,cco illustrations, 1963–73. 10 monographs issued yearly. London, Durrant House, Chiswell Street, E.C.1. All published.
NORDENSKIÖLD (A. E.). *Facsimile Atlas*. Stockholm, 1889. Reprint Kraus, 1961.
—— *Periplus*, 1897. Reprint Franklin, N.Y.
OEHME (H.). *Old European Cities*. London, Thames & Hudson, 1965.
RADFORD (P. J.). Antique Maps, 1971.
RAISZ (Erwin). *General Cartography*. New York, McGraw-Hill, 1948.
SKELTON (R. A.) *et al. County Atlases of the British Isles 1579-1850*. Map Collectors' Circle, 1963–9. Parts I–IV issued 1579–1703.
TOOLEY (R. V.). *Dictionary of Map-makers*. Parts I–XI (A. Powell), 1964–73. All published. London, Map Collectors' Circle, 1964–68 continuing.
—— Bricker & Crone. *Highlights of Cartography*. Amsterdam, Elsivier, 1968.
WATTS (Helen) and Sarah Tyacke (edit.) *My Head is a Map, Essays and Memoirs in Honour of R. V. Tooley*, 1973.
WIEDER (Dr. F. C.). *Monumenta Cartographica*. 5 vols. folio. The Hague, Nijhoff, 1925–33.

3 DECORATIVE SCALE FROM JAILLOT'S MAP OF PALESTINE, 1691

Introduction

IN THE beginning the production of maps was a simple business. The master craftsman drew, engraved, printed, coloured and sold his own maps from his own shop. Francesco Roselli of Florence about 1490 worked in this manner.

As trade increased, apprentices were taken on to do the less important work and surpluses were sold to other dealers, the beginning of the wholesale trade. The main centres for distribution were the great book fairs of Frankfurt, Cologne and Leipzig.

It was mainly the booksellers who developed and frequently financed early map publications. A vastly increased output of prints led to a division of labour. Engraving, generally speaking, became a separate craft, printing another, and colouring a special employment. These were usually controlled by the bookseller, the paymaster. The Italians in the 16th century sold made-up collections of maps, but it was a Fleming, Abraham Ortelius, who organised the map trade. He was the first to issue an atlas uniform in size and contents, with his *Theatrum* published May 22nd, 1570. Theatrum, Theatre, Prospect, or Speculum and Tabularum were terms used to describe a collection of maps. It was Mercator, gifted both in mind and hand, who was the first to use the term Atlas, and either because of its brevity or the renown of Mercator it came to be accepted all over Europe to describe a collection of maps.

Colour is matter of taste. It is the fashion today and was highly esteemed in the past, as is witnessed by such titles as Colourist to the King, the Queen, or the Prince of Wales. The geographer Jaillot is said to have become so enamoured of the delights of maps and colouring, that he abandoned his own profession of sculptor, married the daughter of Nicolas Berey, colourist to the Queen, and produced some of the most magnificent examples of land and sea atlases which he illuminated in gold and colours.

When maps are referred to as being in contemporary colours, it means the original colour applied at the time of printing. Recent colouring is referred to as modern colour.

The printed map has been produced in three main techniques.

1. The design and lettering shown in relief, generally. The woodcut.

2. The design and lettering in intaglio, that is, cut into a metal plate, the ink lying in the grooves. The copperplate.

3. Surface printing. The design and lettering drawn on a prepared surface. The lithograph.

The Woodcut was mostly used in the early days of map production. It had many advantages, the material, wood, was cheap, labour was cheap and its raised surface enabled it to be printed in the same operation with metal type. Its disadvantages were that the material itself was intractable, hence lettering was difficult as was alteration. This was partly overcome by cutting out portions of the woodblock and inserting printed type mounted on wood. As the woodcut map was frequently used with type in book illustration, the wood itself had to be no thicker than the leaded type. Another defect was that if the woodblock was too large it was inclined to warp. The woodcut was used extensively in Germany, Switzerland and to a lesser extent Venice. The woodcut is a line engraving and the ends of the line are square.

Copperplate engraving had the disadvantages of the material itself, copper, being expensive. It needed a stronger and more costly press than the woodcut and any accompanying text had to be printed separately. The preparation of the copper itself was involved, it had to be hammered, ground, honed, polished and heated. The actual engraving was eight times as costly as the metal and it has been estimated that to engrave a square inch was a day's work. The advantages were that copper, a soft metal, made a perfect medium for the burin or engraving tool to move easily over the surface. The engraved line could be varied in depth, dots or strokes added, giving great variety of tone and light and shade that could not be achieved in a woodcut. Intricate designs and the most elaborate lettering were possible. The copperplate could also be altered easily by addition or deletion, the last by knocking out the copper from the back. This greatly increased the useful life of a plate. Copperplate printing started in Italy in the 16th century, gradually spread over all Europe and remained the main medium for map production for three centuries and is still employed in some instances. Copperplate leaves a pointed end.

Surface Printing started in the 19th century with lithography, invented by Alois Senefelder in 1796. He published a textbook in 1818. The advantage of lithography was that the artist or cartographer could draw directly onto the material (the specially prepared stone) without the intervention of any other hand other than the actual printer. This saved time, in fact it was three times faster than the copperplate, and far cheaper, and enabled large scale surveys to be printed when the cost of copperplate would have been prohibitive. Refinements of surface printing were invented, photography, zincography, electrotyping, and the use of metal plates turned into cylinders made possible rotary printing.

All three processes, relief, intaglio and surface, needed a fourth and often a fifth participant, the printer and the publisher. These various occupations are frequently shown on a map distinguished by Latin terms or contractions. Thus

The Cartographer or Draughtsman: Auctore, Auct; Delineavit, Del; Descripsit, Descrip; Invenit, Inv.

The Engraver: Sculpsit, Sculp; Sc. Fecit, fec; and rarely Caelavit.

Printer or Publisher: Apud; excudit, exc; ex officina; formis; sumptibus.

Dating maps has always been difficult. Few maps have a printed date for the obvious reason that few purchasers will buy a map that is out of date, but will accept an undated map.

Even when a map bears a date it is not necessarily the date of publication, but the date of the engraving. For example Speed's maps still bear the date 1610 on copies that were printed years later.

For early atlases it is useful to understand Roman numerals

I – one
V – five
X – ten
L – fifty
C – hundred
D – five hundred
M – one thousand

Any other figure after the above add to get the date, any figure before the above deduct. Thus LX = sixty, XL = forty. There are occasionally some oddities, for example it is impossible to subtract C from L but if they are closed up they make cl. Also ↄ reversed are sometimes used to signify one thousand. Thus CↃCICXLVIIII = 1649.

Discoveries also throw some light on dates. Any map showing the Straits of Magellan must be later than 1520, the date of Magellan's Voyage, and any map that shows Bass's Strait must be after 1797.

STYLES

The *16th century* used architectural motifs and designs taken from the stonemasons and leather workers with interlacing strapwork (Ortelius and Mercator). Colour when used was thick and opaque.

The *17th century* favoured large figured titlepieces with costume figures, fauna, flora and heraldic coats of arms. At the beginning of the century broad side borders with costume figures and top borders of vignette town plans were the vogue. Towards the end of the century, martial and naval motifs became fashionable and contemporary notables in classical attire. Colour was applied in transparent washes, shaded with deeper tones of the same colour or even contrasting colours, such as purple on green (Visscher, Hondius, Blaeu, Jansson, De Wit Jaillot).

The *18th century* was more restrained. Ornament was confined to the titlepiece with light graceful designs modelled on the furniture makers and jewellers' patterns, Rococo work, chinoiseries, with the addition of delicate figures in the style of Watteau and occasional landscapes.

The *19th century* ushered in the Romantic period, the Gothic and Mediæval revival combined with the Nature School, ruins, castles, monastic relics combined with idealised country scenes and rustic virtue provided the inspiration for contemporary design. Ornamentation died out as the century advanced and maps became purely utilitarian.

A partial return to embellishment has occurred in recent times on maps for special purposes, such as trade and travel maps.

There are few hobbies more rewarding than the collecting of early maps, combining as they do both instruction and beauty, presenting a pictorial history of infinite variety.

CHAPTER I

Pre-Christian Geography to Ptolemy

THE Babylonians are credited with making the earliest recorded attempt at a reasoned conception of the universe. It was the Babylonians who introduced the sexagesimal system, dividing the circle of the sky into 360 degrees, the degree into minutes and the minutes into seconds. Likewise the day into hours, minutes and seconds, thus relating the earth to the sky and allowing the former to be plotted in relation to the stars in a constant and proportional manner. According to Herodotus, the Babylonians invented the gnomon, the oldest-known instrument for scientific measurement by means of angles. They regarded the world as round, with an encircling ocean which they named the Briny or Bitter Waters. Outside this circle lay seven islands, possibly representing the seven zones or climates into which the world was divided. These islands formed a bridge to an outer circle or Heavenly Ocean, to which the ancient gods were banished. Later the Babylonians adopted the four cardinal points. In the British Museum is preserved a Babylonian clay tablet on which is drawn a representation of the world dating from the 5th century B.C. The idea of the Babylonian Cosmos also appears on certain gold discs of Germanic workmanship of the second millennium, one of which is preserved in the museum at Hanover. It is from Babylonia also that the most ancient city plans known to us have survived. The oldest known is a cadastral plan of *circa* 2200 B.C., now preserved in the museum at Constantinople; another, of the city of Babylon, dates from the 7th century B.C., and another, of the city of Madakta, in the form of a bas-relief, was made to adorn the walls of the palace of Ashurbanipal, 668–626 B.C.

There are several records and some extant examples of map-making in ancient Egypt. Rameses II, about 1300 B.C., is credited with being the first to attempt to survey the Nile lands, though even before his time Khonsa, the plan-maker, was an important deity. The oldest extant papyrus showing a map is the so-called Turin Papyrus about 1320 B.C., which gives a sketch of an Egyptian gold mine, showing the nature of the country, houses, buildings and entrances to the galleries. The site is unknown.

4 *Babylon World Map, 5th Century*

3

But knowledge of the world as known to the ancients rests primarily on Greek effort. Homer conceived the world as being in the shape of a flat shield surrounded by a wide river, Oceanus. The centre was placed at Delphi, considered the navel of the habitable world. The principle of an outer ocean was reaffirmed by Hesiod, Anaximander, Hecateus and Strabo. The early Greeks had the merit of a true scientific approach to their subject, their errors being in the main due to the poorness of the means at their command. Where they had no knowledge they were content to leave a blank in their maps—a view-point rarely appreciated again till the 19th century.

Anaximander of Miletus, 6th century B.C., is stated to have constructed a map of the world, and Hecateus also of Miletus a little later, about 501 B.C. compiled the first formal geography.

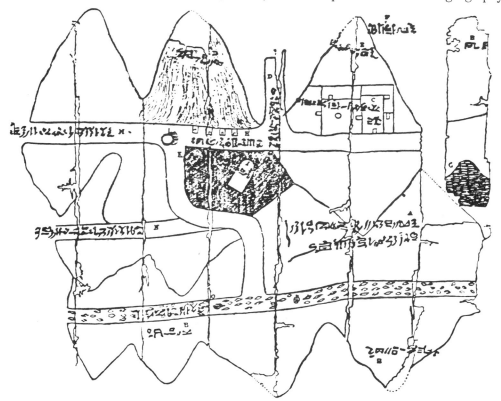

5 *Map of Egyptian Gold Mine*, 1320 B.C.

It was in the form of a Periplus or seaman's guide, giving the number of days' sailing from one place to another. Herodotus, who travelled in Egypt, Babylonia and Persia, collected information on neighbouring lands, and was able to give a considerably enlarged knowledge of the world in his history 444 B.C. The expeditions of Xenophon and Alexander, the embassy of Megasthenes to the Punjaub, the travels of Pytheas to Britain and the Low Countries, the voyages of Nearchus and Hippalus, together with the information brought back from colonies scattered along the shores of the Mediterranean and Black Seas, still further enlarged the fund of information available to the Greek world. This information was summed up by Eratosthenes, librarian at Alexandria about 200 B.C. Eratosthenes, by measuring the shadow of the sun at Alexandria and Syene which he assumed to be on the same meridian, arrived at a remarkably accurate estimate of the circumference of the world. The next two hundred years witnessed the rapid spread of the Roman Empire, and the new facts brought to light were recorded by Strabo 20 B.C.; but though Strabo made additions and corrections to Eratosthenes, he did not alter his main principles.

Every writer on geography has paid tribute to the work of Ptolemy, A.D. 150. He stands like a colossus astride the ancient world, and his influence is still felt to-day. Claudius Ptolemæus, astronomer and geographer, achieved pre-eminence in both these important branches of human knowledge. For fourteen centuries the astronomical theories set forth by Ptolemy in his *Almagest* held undisputed sway, and were only finally dissipated by Newton. His *Geographia* dominated the whole of the Christian and Moslem world for 1,500 years. Ptolemy, living in Alexandria, had access to the greatest library of the time, and was able to consult the works of Eratosthenes, Marinus of Tyre, and all the knowledge painfully gathered in the preceding four hundred years. This he was able to supplement and in part check from personal contact with ship-masters, merchant travellers and other visitors to Alexandria, the hub of trade between east and west. Into all this accumulated data he introduced an order or system that has since been followed by geographers in all ages. It was Ptolemy who introduced the method and names of latitude and longitude. His work had two main faults: by his method of measurement and choice of the Canary Islands for his prime meridian he greatly overestimated the length of the land surface eastward from this line, and consequently reduced the gap, presumed to be water, lying between Europe and Asia. And whereas the early Greeks had been content to leave blanks in their maps where knowledge ceased, Ptolemy filled in the blanks with theoretical conceptions. This would not have mattered so much in a lesser man, but so great was the reputation of Ptolemy that his theories assumed an equal validity with his undoubted facts, and were not seriously questioned for 1,500 years. In spite of its defects, Ptolemy's work was far in advance of previous efforts in this field, and it is only necessary to look at the poverty of independent geographical work of the nations of western Europe right up to the 16th century to realise in part the greatness of the Alexandrian. (7 and 8) The most notable Roman remain is the Peutinger Table (route map), possibly by Castorius (second half of the 4th century), discovered by Peutinger in the 16th century and named after him. (6)

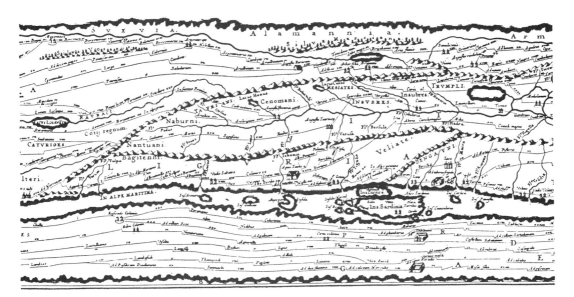

6 *Part of a Roman Route Map (Peutinger Table)*

LIST OF VARIOUS EDITIONS OF PTOLEMY'S GEOGRAPHY

1475 (folio) *Vicenza. Hermanus Levilapis [Hermann Lichtenstein].*
First edition. Issued without maps. Translated by J. Angelus, edited by Angelus Vadius and Bernardus Picardus. 144 leaves.

(1477) (folio) *Bologna. 26 copper-plate maps.*
First engraved atlas in the world, the rarest and most valuable edition of Ptolemy. Translated by J. Angelus, with corrections by Beroaldus and others, edited by Angelus Vadius. 61 leaves, 26 maps, first leaf a blank. The atlas is misdated 1462.

1478 (folio) *Rome. Arnoldus Buckinck. 27 copper-plate maps.*
Translated by J. Angelus, emendations by D. Calderinus. 70 leaves without title, signatures or pagination, last leaf blank. 27 maps.

(1482) (folio) *Florence. Nicolo Todescho. 31 copper-plate maps.*
The first atlas to attempt the introduction of modern geography. Four new maps—France, Italy, Spain and Palestine—being based on contemporary knowledge. The text is a metrical paraphrase by Francesco Berlinghieri, and is the First Edition in Italian. It is also the only edition with maps printed on the original projection with equidistant parallels or meridians. Nordenskiöld dates this edition 1478, giving it priority over the Roman edition.

1482 (folio) *Ulm. Leonardus Hol. 32 wood-cut maps.*
The first edition to be printed in Germany, the first with wood-cut maps. The map of the world is the first to show contemporary discoveries, and the first map to bear the name of its engraver, Johannes Schnitzer de Armssheim. Five of the maps modern. One, the first to show a printed representation of Greenland. Translated by J. Angelus, edited by Nicolaus Germanus, who drastically revised the maps. 70 leaves, last blank, 32 maps. Some copies were printed on vellum, and there are variants of the text.

1486 (folio) *Ulm. Johann Reger. 32 wood-cut maps.*
B.M. catalogue 1895 states the maps and type are the same as those of edition of 1482. I have, however, noted additions to the map of Germany in the 1486 edition. 140 leaves (117 and 124 blank), 32 maps.

1490 (folio) *Rome. Petrus de Turre. 27 copper-plate maps.*
Second Rome edition with same maps as 1478 edition. 120 leaves (1, 36, 37, 98 and 120 blank), 27 maps.

1507 (folio) *Rome. Bernardus Venetus de Vitalibus. 33 copper-plate maps.*
27 of the maps are re-issues of the Rome editions of 1478 and 1490. The other 6 maps—northern Europe, Spain, France, Poland, Italy and the Holy Land—are based on contemporary knowledge. 127 leaves, 33 maps.

1508 (folio) *Rome. Bernardus Venetus de Vitalibus. 34 copper-plate maps.*
A re-issue of the preceding, but with a new title-page, an account of the New World by Marcus Beneventanus, and a new map of the world by Ruysch, "Nova Tabula." This is the first map in any edition of Ptolemy to show the New World. 141 leaves, 34 maps.

1511 (folio) *Venice. Jacobus Pentius de Leucho. 28 woodcut maps.*
The heart-shaped map of the world is the first in an edition of Ptolemy to show a printed delineation of part of the North American continent. Translated by Angelius, with annotations by Sylvanus. Some copies are printed on vellum.

1513 (folio) *Strassburg. Johannes Schott. 47 wood-cut maps.*
The most important edition of Ptolemy, containing the 27 maps of the ancient world and 20 maps based on contemporary knowledge, under the superintendence of Martin Waldseemüller. Includes the Tabula Terra Nova, the first map specifically devoted to the delineation of the New World. The greatly increased number of "modern maps" makes this in effect the first modern atlas.

1514 (folio) *Nuremberg. Johannes Stuchs. No maps.* New translation by Joannes Werner.

1520 (folio) *Strassburg. Johannes Schott. 47 wood-cut maps.*
Maps as in 1513 edition from the same blocks, except Switzerland.

1522 (folio) *Strassburg. Joannes Gruninger. 50 wood-cut maps.*
Reduced versions of 1513 edition plus three new maps compiled for this edition, viz. World by Laurent Frisius; Tab. Mod. Orientalis; Tab. sup Indiae et Tartariae Majoris.

1525 (folio) *Strassburg. Joannes Gruninger. 50 wood-cut maps.*
Maps, with the exception of Asia V, printed from the same blocks as 1522 edition, and like them almost unaltered copies on a reduced scale of the maps of the 1513 edition.

1533 (quarto) *Basle. Hieronymus Froben.*
First edition with Greek text. No maps. Edited Erasmus.

1533 (quarto) *Ingoldstadt. No maps.*

1535 (folio) *Lyons. Melchior and Gaspar Treschel. 50 wood-cut maps.*
Maps as in 1525 edition. Edited Servetus. Ordered to be burned by Calvin.

1540 (octavo) *Cologne. Joannes Ruremundanus. No maps.*

1540 (folio) *Basle. Henricus Petri. 48 wood-cut maps.*
Edited by Seb. Münster, who is the first to quote his authorities for the modern maps.

1541 (folio) *Vienne. Dauphiné. Gaspar Treschel. 50 wood-cut maps.*
Maps as in 1535 edition.

1542 (folio) *Basle. Henricus Petri. 48 wood-cut maps.*
A re-issue of the 1540 Basle edition, with slight variations. Wood-cut designs on back of map attributed to Holbein.

1545 (folio) *Basle. Henricus Petri. 54 wood-cut maps.*

1546 (octavo) *Paris. C. Wechel. Greek Text.*

1548 (octavo) *Venice. Nicolo Bascarini. 60 copper-plate maps.*
Apart from the 1480 metrical edition of Berlinghieri, this is the first edition with Italian text. Translated by Mattioli. A notable edition, the maps being designed by Jacobo Gastaldo after those of Munster, but with important additions.

1552 (folio) *Basle. Henricus Petri. 54 wood-cut maps.*
Maps as in 1545 Basle edition.

1561 (quarto) *Venice. Vincenzo Valgrisi. 64 copper-plate maps.*
Italian text. The maps are enlargements of Gastaldo's maps of 1548 edition. The first atlas (though not map) to show the world in two hemispheres.

1562 (quarto) *Venice. Vincenzo Valgrisi. 64 copper-plate maps.*
Maps same as in 1561 edition.

1564 (quarto) *Venice. Giordano Ziletti. 64 copper-plate maps.*
Italian text.

1564 (quarto) *Venice. Giordano Ziletti. 64 copper-plate maps.*
Latin text.

1574 (quarto) *Venice. Giordano Ziletti. 65 copper-plate maps.*
Italian text. Maps same as in editions of 1561, 1562 and 1564, but with addition of new map of Rome.

1578 (folio) *Cologne. Godefridus Kempen. 28 copper-plate maps.*
Of importance as being the first edition with maps compiled by Mercator for Ptolemy's *Geography*.

1584 (folio) *Cologne. Godefridus Kempen. 28 copper-plate maps.*
Maps as in preceding edition.

1596 (quarto) *Venice. Heirs of Simone Galignani de Karera. 64 copper-plate maps.*
A new edition with new maps (by H. Porro).

1597 (quarto) *Cologne. Petrus Keschedt. 64 copper-plate maps.*
Maps as in 1596 edition.

1597–8 (folio) *Venice. G. B. and G. Galignani. 64 copper-plate maps.*

1598–9 (quarto) *Venice. Heirs of Melchior Sessa. 69 copper-plate maps.*
Italian text. Maps as in editions of 1561, 1562, 1564 and 1574 re-worked.

1602 (folio) *34 copper-plate maps. Dusseldorf.*

1605 (folio) *Amsterdam (also at Frankfurt). Cornelius Nicolaus and Jodocus Hondius. 28 copper-plate maps.*
The first edition to have Greek and Latin text together. The maps are the same as in editions of 1578 and 1584.

1608 (quarto) *Cologne. Petrus Keschedt. 64 copper-plate maps.*
Maps as in editions of 1596 and 1597.

1617 (quarto) *Arnheim. Joannes Jansson. 64 copper-plate maps.*
Maps as in editions of 1596, 1597 and 1608.

1618–19 (folio) *Leyden. Elzevier, pub. Amsterdam Hondius. 47 copper-plate maps.*
Greek and Latin text in parallel columns. The maps taken from Mercator's editions of 1578, 1584 and 1605, and from various editions of Ortelius's *Theatrum.*

1621 (folio) *Padua. Brothers Galignani. 64 copper-plate maps.*
Maps as in edition 1597–8.

1695 (folio) *Franeker & Utrecht. 28 copper-plate maps.*
No text. A re-issue of Mercator's Maps, 1578.

1698 (folio) Re-issue of preceding.

1704 (folio) *Amsterdam & Utrecht. 28 copper-plate maps.*
Re-issue of editions of 1695 and 1698.

1730 (folio) *Amsterdam. 28 copper-plate maps.*
Re-issue of editions of 1695, 1698 and 1704.

1828 (quarto) *Paris.*
Greek and French text. Translation by Abbé Halma.

1838–45 *Essendiae Fasc. 1–6 only.*

1843–45 *Lipsiae. Tauchnitz. Greek text.*

1867 *Paris. Didot. Greek text.*

1883 *Paris. Vol. I only.*

AUTHORITIES

AIRENTI (G.). *Osservazioni intorno all' opinione di G. Meermann e di altri scrittori sopra la tavola Peutingeriana,* Roma, 1809.

BEAZLEY (C. RAYMOND). *Dawn of Modern Geography,* 3 vols., 1897–1906.

BLAZQUEZ (A.). *Pyteas de Marsella,* Madrid, 1913.

BONACKER (W.). "The Egyptian 'Book of the Two Ways'" (*Imago Mundi VII,* 1950).

BRUGSCH (H.). *Die Geographie des alten Ägyptens,* Leipzig, 1857.

BUNBURY (E. H.). *History of Ancient Geography,* 2 vols., 1879.

COLUMBA (G. M.). *Eratostene e la misurazione del meridiano terrestre,* Palermo, 1895.

EAMES (W.). *List of editions of Ptolemy's Geography, 1475–1730,* New York, 1886 (limited to 50 copies).

ECKERT (M.). *Die Entwicklung der kartographischen Darstellung von Stadtslandschaften,* 1930.

GAMBIA (R.). *Osservazioni sur la edizione della Geographia di Tolomeo fatta in Bologna.*

GOSSELIN (P. F.). *Géographie des Grecs analysée,* 1790.

HEIDEL (W. A.). *The Frame of the Ancient Greek Maps,* N.Y., 1937.

HERRMANN (A.). *Marinus von Tyrus* (Peterm. Mitt.), 1930.

JACOBS (J.). *Story of Geographical Discovery,* Newnes, 1899.

LYNAM (E. W. O'F.). *The First Engraved Atlas of the World, The Cosmographia of Claudius Ptolemaeus Bologna 1477,* Jenkintown, 1941.

—— *Notes on Cosmographia of Ptolemy Bologna 1477,* 1948.

NORDENSKIÖLD (A. E.). *Facsimile Atlas,* 1889.

PINZA (G.). *Due Cosmografie una egizia ed una sinaitica in due piatti di bronzo trovati a Nimrud,* Roma, 1913.

RYLANDS (T. G.). *Geography of Ptolemy Elucidated,* 1893.

STEVENS (Henry). *Ptolemy's Geography,* 1908.

TOZER (Rev. H. F.). *History of Ancient Geography,* 1897. Second Edition, 1935.

—— *Lectures on the Geography of Greece,* 1873.

UNGER (E.). "From Cosmos Picture to World Map" (*Imago Mundi II,* 1937).

—— "Ancient Babylonian Maps and Plans" (*Antiquity,* 1935).

WARMINGTON (E. H.). *Greek Geography,* 1934.

WINSOR (J.). *Bibliography of Ptolemy's Geography,* Cambridge, Mass., 1884.

YOUSSOUF KAMAL (Prince). *Monumenta Cartographica Africae et Aegypti,* 1926.

—— *Quelques Eclaircissements épars sur mes Monumenta Cartographica Africae et Aegypti,* 4to, 1935.

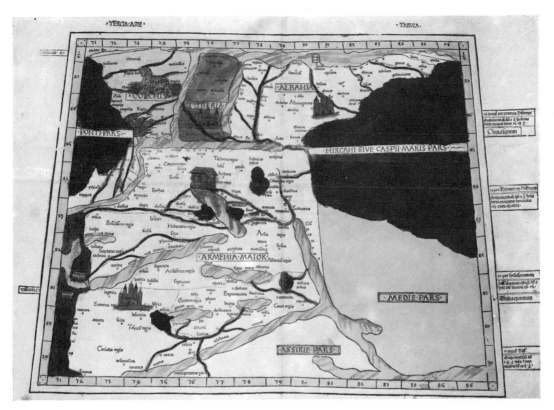

7 PTOLEMY'S *ARMENIA*, WITH NOAH'S ARK RESTING ON MOUNT ARARAT
($19\frac{1}{2}'' \times 14''$)

8   PTOLEMY'S MAP OF THE WORLD ($21\frac{1}{2}'' \times 13\frac{3}{4}''$)

from the *Cosmographia Bologna*, 1477. The first engraved atlas of the world. (*B.M. Maps C1e 12*)

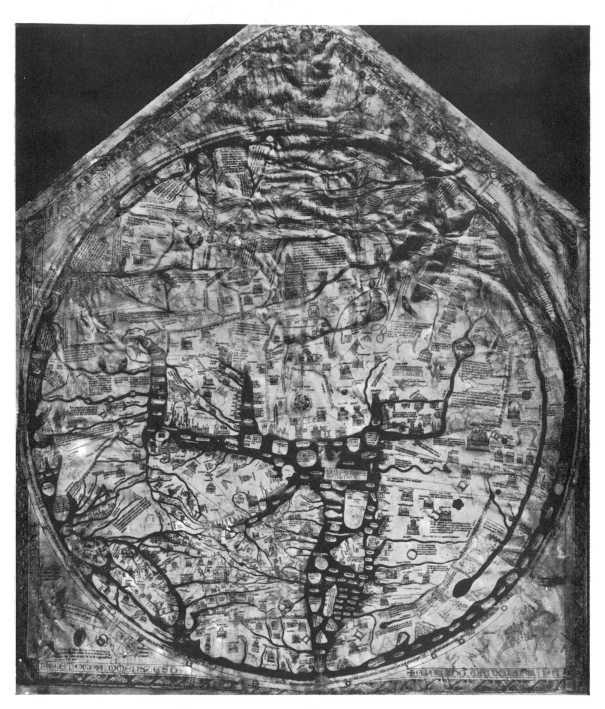

9 HEREFORD MAPPA MUNDI, *c.* 1280

CHAPTER II

The Arabs and Mediæval Europe

UPON the dissolution of the Roman Empire and the consequent separation of the nations into small units, science (or the greater knowledge, as distinct from local information) was submerged. The gradual rise of a central unifying force, the Arab Empire, once it was stabilised, led, if not to a general advance in cartographic knowledge, at least to a revival of interest in the subject. The Arabs became the intellectual heirs of the Greeks, and it was Moslem thought that stimulated and exercised the best minds in the west. From the 7th to the 12th centuries, science and geographical knowledge came to Europe primarily through the great Arab centres of learning—Baghdad, Cordova and Damascus. The philosophical, mathematical and astronomical speculations of the schools of Bologna, Rome, Oxford and Paris in the 13th century were, in the main, reflections of Moslem study. Thus it is to the Arabs we owe the maintenance of the link between the ancient learning of classic time and the revival of western science in the Renaissance.

Cartographically the Arabs adopted the Ptolemaic theory of the Terrestrial World, with a land-locked southern sea, rejecting the more correct earlier Greek conception of an encircling ocean. In Moslem eyes Ptolemy had the advantage of being the most symmetrical of the Greeks, and as the voyages of Arab subjects gradually disproved many of his theories, it should not be overlooked entirely that it was politically advantageous to the Arabs to stress the only possible route to the east as the landward one under their control. Hence for various reasons Arabian scholars postulated a limitless Atlantic Ocean, "the Green Sea of Darkness," in the west, and to the south a great country "uninhabitable by reason of the heat."

Ptolemy's *Geography* was translated into Arabic in the 8th century, probably from Syriac texts. It was not long before additions were made to the work of the great Alexandrian by Al Kharizmi, who had studied trigonometry in India. Alfergany was the first Arab to write on the astrolabe, and in the 9th century Ibn Khordabeh described the trade routes between Europe and Asia. In the 10th century Arab science reached its apex under Albateny and Massoudy. The former made important astronomical discoveries, and the latter was the first to depart from the strict Ptolemaic tradition and veer towards Strabo's cosmography, for though he preserved an impenetrable land-mass in the south, he opened the Indian Ocean in the east. On the other hand, he decreased still further the circumference of the world than had his predecessors.

In the 12th century knowledge of the Moslem world was summed up by Alberouny in the east and Edrisi in the west. The celebrated world map of Ibn Haukel was based on the work of Alberouny. It was Edrisi, however, who gave the greatest lustre to Arab science in the eyes of the western world. Born at Ceuta in 1099, he finally settled at Palermo as geographer to the Norman court of Roger of Sicily. Edrisi's conception of the world was a blending of pagan and Christian learning, a combination of the theories of Eratosthenes, Ptolemy and Strabo, modified by the work of his Arab predecessors, and possibly a partial knowledge of northern lands from his hosts. Edrisi divided his world into the seven climates of the early Greeks, kept the Indian Ocean land-locked on the south, but open in the east, with an outer encircling ocean.

On the whole, Arab contribution to cartography was disappointing. It was a period of great scholastic speculation rather than practical advancement. This is the more surprising considering the real advance made in topographical knowledge by the extensive travels of

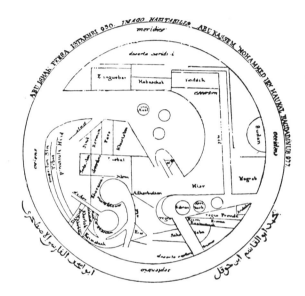

10 *Arabian Map of the World*

Arab subjects, ranging from Spain in the west to China in the east, and their penetration north-wards into Russia, and southwards down the east coast of Africa. For example, Solyman in the 8th, Benjamin of Tudela in the 12th, Al Heravy in the 13th and Ibn Batuta in the 14th centuries, probably had a knowledge of a far greater portion of the globe than any European of their time. Yet in spite of their inherited and acquired knowledge and the undoubted ability of their savants, the Arabs did not acquire the technical ability to reproduce graphically their accumulated facts, and consequently did not make any serious attempt to correct the postulates of the earlier Greek geographers.

To turn to the western world, the most important event in the Middle Ages was the rise of the Church to secular power. This in its early stages was of inestimable value to science, giving peace and, to those who so desired, the chance to study within the fold of the Church. Popular appreciation of the Middle Ages has almost invariably been unjust. If civilisation is the art of living up to the resources of a period, the Middle Ages should rank high. Few later figures can compare in profundity of religious feeling with St. Augustine, in humanity with St. Francis, in comprehensive erudition with Adelard of Bath, Archbishop Siegeric, Roger Bacon, Albertus Magnus, Gerard of Cremona and Aquinas; or in speculative skill with Pierre d'Ailly. The ignorance of the Middle Ages, so often inferred by later commentators, comes somewhat ungenerously from the specialist, who frequently criticises but one facet of their

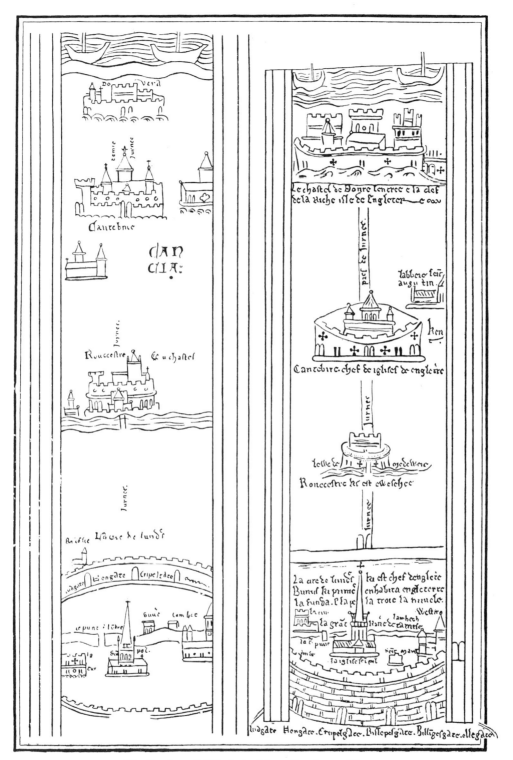

11 *Section of Pilgrim's Map, London to Dover*

activities, and in cartography this applies with particular force, due partly to the paucity of extant records—their wide dispersal and consequent difficulty of research—but also to a misapprehension of the motives actuating mediæval thought.

It is unwise to assume that mediæval scholars were as ignorant as their maps would imply. Their aim might be, and probably was, symbolic and moral rather than utilitarian. Ecclesiastics showed their mappæmundi or world maps for the edification of the pious, usually to illustrate an historical or encyclopædic work, or as a pictorial representation of Church dogma. These mappæmundi were scholastic conceptions of a theoretical universe, and were not intended for, or used by, travellers or traders, who had their own guides and charts, the last being remarkably accurate in the latter part of the period.

Thus mediæval maps may be roughly divided into the symbolic and theoretical and the practical. The first, mainly the province of the ecclesiastic, have been far more widely publicised than the last group, which consists of nautical charts entirely in the hands of laymen. These were, all things considered, remarkably accurate. Steady research in recent years has brought to light no less than six hundred examples of mediæval mappæmundi dating from the 8th to the 15th centuries.

12 *Mediæval Map*

The symbolic and theoretical maps are sometimes subdivided into the Oecumenical (habitable world) and Hemispherical (representing the whole hemisphere). The simplest form of mediæval conception of the world was diagrammatic, the three continents of the ancient world being symbolised by the placing of a T within an O, with the east at the top. These rudimentary diagrams are known as TO maps. Though pagan in origin, they conveniently fitted in with Church teaching, the earth divided among the three sons of Noah, and some of these tripartite maps are lettered Shem, Ham and Japhet respectively. Occasionally, in some early examples, the idea was presented in the form of a triangle or square—a Y written within an O or a V within a square. With the course of time these maps were elaborated, though their own basic principle remained the same. About a hundred examples of TO maps are so far known, occurring in MSS. from the 9th to the 13th centuries, the most ancient illustrating the works of St. Isidore of Seville and Paulus Orosius, e.g. the Albi world map of the 8th century (the oldest-known geographical monument of western Europe, dating from the time of the Venerable Bede), a 9th-century MS. at Strassburg, a 10th-century MS. of Lambert, Canon of St. Omer, preserved in Ghent, a 12th-century MS. of William de Conches at Cambridge, another in Brussels (*Codex Guidonis*), and a 13th-century example at Metz. An elaboration occurs in the so-called Sallust maps, particular attention being paid to Africa. These maps were originally drawn to illustrate the MS. of Bellum Jugurthinum. A 13th-century example, preserved in the Ambrosian Library at Milan, has several subdivisions, Egypt and Spain being prominently shown, and with Jerusalem as a castellated tower.

Another popular representation, more mathematical in conception, accepted the sphericity of the world, and portrayed a single hemisphere divided into five zones or seven climates of the habitable world, with an encircling ocean, frequently intersected by two oceans at right angles, a double oceanic theory, one equatorial, the other passing through the Poles. These are called the hemisphere or zone maps, and they were extensively used to illustrate the works of Macrobius and Martianus Capella. The climatic theory of the earth was taken up and developed by the Arab scholars, and examples also occur in the MS. works of Sacrobosco (John of Holywood) and Petrus Alliacus (Cardinal Pierre d'Ailly).

The following are some of the better-known examples of map-making from the time of Ptolemy to the close of the 15th century:

c. A.D. 550: Madaba Mosaic

The most ancient example of Christian topography known to exist is the Madaba Mosaic, dating from the 6th century. It shows Palestine and parts of Arabia, Egypt and the Mediterranean. Originally on a large scale, it measured roughly 50 by 20 feet, much of it has been destroyed. The most important part remaining contains a perspective plan of Jerusalem. It is the earliest example of a map with the east at the top, and the only known example of Byzantine cartography. The mosaic was discovered in 1896 on the floor of a church near Madaba, about 15 miles from the Dead Sea.

6th Century: Cosmas Indicopleustes

Cosmas, an Egyptian traveller and monk, devised a map of the world "evolved out of Holy Scripture," which he interpreted very literally. He denounced the Greek theory of the sphericity of the world as contrary to Holy Writ, delineating his world as a flat parallelogram, with Jerusalem in the centre. His work was individual and his following small, even in his own day, and his Christian Topography soon fell into oblivion.

A.D. 670: Plan of Jerusalem

Arculf gave a description of his visit to the Holy Land to Adamnan, abbot of Iona, for which he drew a map. This map shows the circuit of the walls, and the various gates are named. The east is at the top.

7th Century

Isodorus of Seville world map.

8th Century: Albi Map

A world map to illustrate the cosmography of Honorius and Orosius. It is the oldest geographical monument of western Europe, dating from the time of the Venerable Bede and Charles Martel.

8th Century: Beatus

Beatus, a Benedictine monk of Valcovado, in Spain, compiled a world map, the original of which has been lost. Evidently popular in the Middle Ages, ten extant examples, supposed copies of the lost original, are known, viz. Ashmolean, A.D. 970; St. Sever, A.D. 1030; Valladolid, A.D. 1035; Madrid, A.D. 1047; London, A.D. 1109; Paris, A.D. 1150; Turin, 12th century; Gerona, 12th century; Osma, A.D. 1203; Paris, 1250.

A.D. 742–814

It is recorded that Charlemagne had three silver tables constructed on which maps were drawn showing Rome, Constantinople and the world.

10th Century: Anglo-Saxon

Map of the world drawn about A.D. 1000. Found in one of the Cotton MSS. in the British Museum. Its eastern limit is the Persian Gulf and Red Sea, and the Caspian is open at north, but it is notable for its rendering of the north-west and British Isles, and is far superior to any other map of the time, and even those of a later period.

11th Century: Turin Map

c. A.D. 1100: Situs Hierusalem

Several examples of this plan, perspective in form, are known. Less correct than the Madaba Mosaic. In the copy preserved in Brussels, Jerusalem is shown within a circle, in the Montpellier copy within a square.

c. 1100: Henry of Mainz.

A.D. 1119: Lambert, Canon of St. Omer

Compiled a map of Europe and a TO map. Of particular interest for northern regions.

13th Century: Sacrobusco

Joannes de Sacrobosco (or Holywood), teacher of mathematics and astrology at the University of Paris in the first part of the 13th century, composed his *Opusculum Sphericum*. Extremely popular. Sixty to seventy versions appeared from the 13th to 15th centuries. His world map is hemispherical, the habitable half of the world divided into seven zones or climates. This is succeeded by an uninhabitable zone, and beyond, a temperate zone again.

13th Century: Matthew Paris of St. Albans

Matthew Paris, monk and historian, compiled his *Chronica Maiora* and *Historia Anglorum*. Maps were compiled for these in the scriptorium of St. Albans Abbey, one it is said by John of Wallingford, who died in 1258. The map of England is the earliest-known detailed map. Four examples have survived, and have been reproduced by the British Museum.

13th Century: Psalter Map

In the British Museum there is a circular world map, sometimes called cartwheel, a typical example of religious geography.

1275–1317: Hereford Mappamundi

One of the most famous of mediæval maps, supposed to have been compiled by Richard of Haldingham. It is preserved in Hereford Cathedral. Circular in form, with the east at the top, it shows representations of Paradise, Last Judgment, Pillars of Hercules and numerous biblical and other legends. Jerusalem is in the centre. The British Isles are shown on the edge of the map, bent round to fit in the circle, Scotland being separated from England. (9)

1284: Ebstorf Mappamundi (now destroyed)

Similar to the Hereford map.

1306–21: Marino Sanudo the Elder

Composed a work, *Liber secretorum fidelium crucis*, which was accompanied by maps. Several manuscripts are known of this work, the most complete containing ten maps (Black Sea, Mediterranean, Western Europe, World, Palestine, and plans of Jerusalem, Ptolemais and Antioch). Very fine representation of Africa. Maps attributed to Petrus Vesconte.

14th Century: Gough's Map (Bodleian)

An anonymous map of Great Britain of the time of Edward III was first described by Gough in his *Antiquities*, 1780. It shows a great advance in place-names and shape of England over previous work, but Scotland and Ireland are still inadequately represented.

14th Century: Ranulph Higden

The *Polychronicon* of Ranulph Higden contains an oval world map, with Adam and Eve at the top. Several manuscripts are known, three variants being given in Santarem's atlas.

c. 1400: Este World Map

A.D. 1410: Cardinal Pierre d'Ailly

World map in his *Ymago Mundi*. Hemispherical, the whole of the habitable world within the northern half. The various countries are indicated by name only, with their relative position one to another. One of the earliest to place the north at the top of the map. A printed version appeared in Louvain in 1483

A.D. 1440: Buondelmonte

Supplied a perspective plan of Constantinople in his manuscript island atlas *De Insulis Cyclades*. Shows in detail buildings, palaces, St. Sophia, Hippodrome, etc.

A.D. 1452 : Leardo World Map

A.D. 1453: Borgian World Map

Engraved on metal. Preserved in Cardinal Stephen Borgia's Museum at Velletri.

A.D. 1459: Fra Mauro's World Map

Considered one of the most important cartographical works of the 15th century. Full of detail and legends, it forms a "mediæval cosmography of no small extent." Part of the map was based on portolan sources, and progress is shown in the delineation of Scandinavia and Africa. The Asiatic portion is based on Marco Polo.

Almost contemporaneous with the religious scholastic representations of the world, an entirely different type of map was being produced, the portolano.

The portolans were charts made by seamen for seamen. They were practical and utilitarian, at first almost entirely the work of Italian and Catalan draughtsmen, later of Portuguese and other nationalities.

The typical portolano had no graduation, but instead a network of loxodromes or rhumb lines, that is, straight lines in the direction of different winds, their points of intersection on the later examples being formed into compass roses, many of great beauty. These rhumb lines, however, bore no relation to the actual construction of the map itself, being added after the drawing of the chart and varying from one portolan to another.

In other respects the portolan draughtsmen were extremely conservative, not only the map itself being copied with minute accuracy, but a systematic rule for colouring being adhered to, black, red, green, blue, yellow, gold and silver having their respective places; and the same on the Catalan as on the Italian examples.

At first, dealing solely with the Mediterranean and Black Sea area, they were concerned only with a delineation of the coast and its salient features, no details being given of the interior. Successive discoveries were added gradually to the main chart, first west and north-west Europe, then Africa, and lastly America, each succeeding type, once firmly established, being copied as faithfully as its predecessors, the central core remaining the same.

These portolans were remarkably accurate from the very beginning for the Mediterranean area, and the outer bounds, though rudimentary at first, were continuously if slowly improved, and far in advance of the map of the geographers who generally ignored their work up to the introduction of the "Tabula Moderna" in the various editions of Ptolemy.

The oldest recorded example of a portolan, the Carte Pisane, dates from the beginning of the 14th century, the earliest-dated example being that by Petrus Vesconte, 1311. But these must evidently have been based on still earlier models. Certainly written records of chart-making go back to a far earlier period, and it has even been suggested that the prototype derives from the charts of Marinus of Tyre, through Byzantine channels.

Many of these portolans are in atlas form, containing from 4 to 10 or 12 charts. For example, 3 atlases besides separate charts are known of the work of Petrus Vesconte, that of A.D. 1320 containing a world map, 6 charts of the normal portolan area, a map of Palestine and plans of Jerusalem and Ptolemais.

An elaborate and famous example is the Catalan portolano of Angelino Dulcert, A.D. 1339. It includes the Canary Isles, and shows the junction of the Blue and White Nile. The compass is divided into 32 points for the first time in this map. A more remarkable rendering of Africa is given in the so-called Laurentian Portolano, A.D. 1351, showing the Gulf of Guinea, and suggesting the possibility of rounding the continent long before the Portuguese achieved the feat. It is an atlas of 8 plates, including special charts of the Adriatic, Caspian and the Archipelago.

The Catalan World Map of A.D. 1375 was composed for Charles V of France. It is on parchment, brilliantly coloured, mounted on leaves to fold like a screen. It is preserved in Paris. Nordenskiöld calls it "the most comprehensive cartographic work of the 14th century." This western part is a portolan, the eastern part based on Ptolemy, with additions from the travels of Marco Polo.

One of the most prolific of portolan makers was Grazioso Benincasa, no less than 25 of his works, dating from A.D. 1435 to 1482, being known, 10 of them being in atlas form. Contemporary with Benincasa was Andrea Bianco, one of whose charts, that of A.D. 1448, is of interest as being drawn in England.

One of the most precious portolan atlases was that acquired by the British Museum, containing 35 charts, being early copies (circa 1500) of charts by Roselli, Benincasa, Zuan de Napoli, Francesco Becaro, Cexano, Pasqualin, etc. The Martellus Portolano of 1492, also in the British Museum, is of interest as being one of the earliest to depict the Cape, the result of the voyage of Diaz in 1486.

The art of chart-making was frequently a family affair, descending from father to son, as for example in the Olives, Freducci and Homem families. Of the later chart-makers one of the most distinguished was Battista Agnese, A.D. 1527–64; not so much on account of new geographical information, but for the beauty and technical perfection of his work. The finest examples of the 16th century were brilliant works of art: gone was the austerity of the early masters, no longer was the portolan purely utilitarian, a tool for a seaman, but a present fit for a prince, and in many cases composed specially for them. In form it remained the same, accurate delineation of the coast-line as far as knowledge allowed being still the primary principle; but the interior and the blank spaces of the sea blossomed into a pictorial and heraldic geography as gay and vivid, and almost as full of fantasy and legend, as a mediæval mappamundi.

List of some of the better-known Portolan Makers

14th Century

Early 14th Carte Pisane	1306 Giovanni da Carignano	1351 Anon. (Laurentian)
Early 14th (Tammar Luxoro)	1311–27 Petrus Vesconte	1367–73 Francesco Pizigano
Early 14th Marino Sanudo the Elder	1321 Perrinus Vesconte	1375 Anon. (Catalan Atlas)
	1330 Angellino Dulcert	1385 Guillelmus Soleri

15th Century

1408 G. Pasqualini	1448–52 Giovanni Leardo	1492 Henricus Martellus
1413 Meca de Villadestes	1456–89 Petrus Roselli	1494 Giovanni Georjio
1421 Franciscus de Cesanes	1461–82 Grazioso Benincasa	1497 Conte Hectomanni Freducci
1422–43 Jachobus de Giroldis	1470 Nicolaus de Nicolo	1497 Jehuda Ben Zara
1426–35 Batista Becharius	1476–90 Andreas Benincasa	15th cent. Pietro Rosso
1427 Claudius Clavus	1480 Albino de Canepa	15th cent. Buondelmonte
1430 Chola de Briaticho	1480 (Nicolo Fiorin)	15th cent. Francesco Cexano
1436–48 Andrea Bianco	1482 Jacme Bertran	15th cent. Aluixe Cexano
1439–77 Gabriel de Vallsecha	1487 Nicholas Mare	15th cent. Nicolo de Pasqualin
1455 Bartolomeus de Paretto	1489 Zuan de Napoli	

16th and 17th Centuries

1497–1556 Freducci Family of Ancona: Conte Freducci, Angelus Eufredatus	1537–65 Georgio Calapoda	1593–1607 Vincentius Demetrius Voltius
	1542 Rocco dall Olmo	1596 Bartolomeo Crescentio
	1544 Francesco Lodesano	1600 Thomas Lupo
1504–86 Maiolo or Maggiolo Family: Vesconte, Jacobus, Baldassore, Giovanni	1550 Jaume Olives	1607 Andrea Rios
	1553–90 Matheus Prunes	1612 Nicolas Reynolds of England
	1557–75 Diego Homem	
	1557–90 Antonio Millo	1612–30 Alvise Gramolin
1511 Salvat de Pilestrina	1558 Bastian Lopez	1613 Mario Cartaro
1514 Battista Genovese	1562–69 Paulo Forlani	1613–33 Gio. Francesco Monno
1520 Johannes Xenodochos of Corfu	1564–86 Johannes Martines	
	1568 Domingo Olives	1615 Sebastian Condina
1520–88 Jacobus Russus	1570 Bartolomeo Bonomini	1618 Petrus Cornetus
1524–30 Francisco Rodrigues	1571 Giulio Cesaro Petrucci	1629 J. F. Mon
1527–64 Battista Agnese	1578–92 Jacobus Scottus	1645–50 Alberto di Stefano
1530–80 Dominicus Vigliarolus	1581 Mateus Gruisco	1646 Nicolo Guidalotto
	1589 Domingo de Villaroel	1650 Gio. Battista Cavallini
1532–88 Bartolomeo Olives	1590 Muhammed Raus	1651 Pietro Giovanni Prunes
	1590 Jaimes Dossaiga	1663–69 J. F. Roussin
	1590 Augustinus Russinus	
	1592 Carlo da Corte	

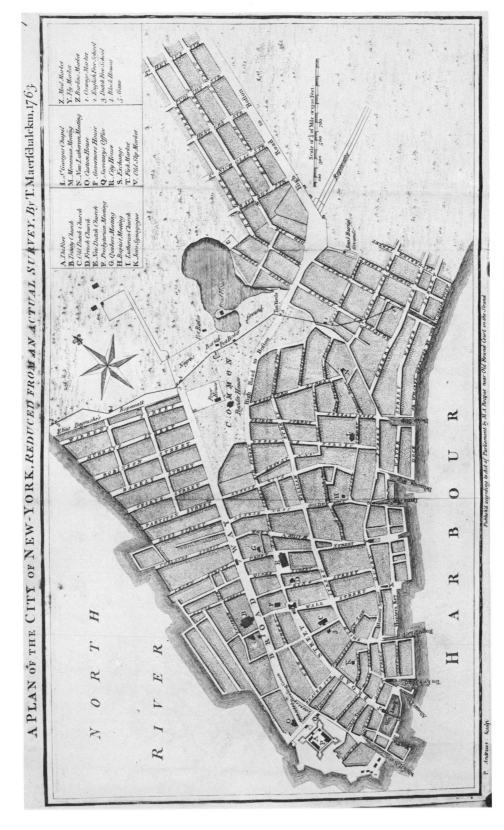

13 MAERSCHALCKM'S CITY OF NEW YORK, 1763

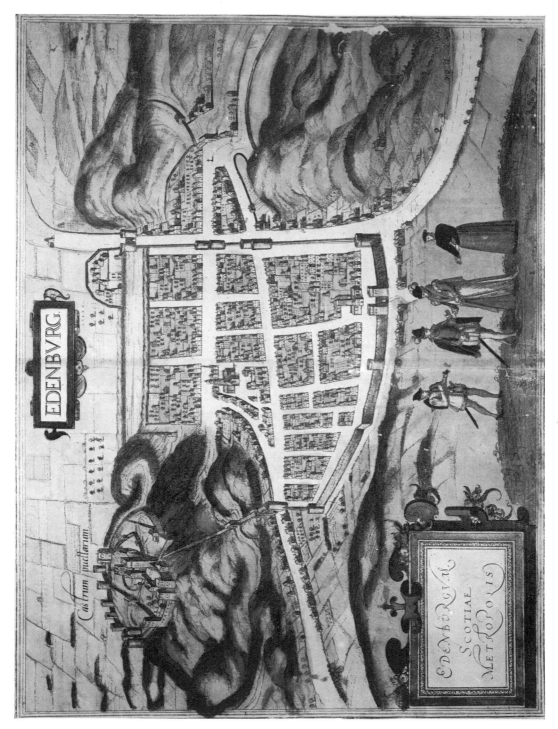

14 BRAUN AND HOGENBERG'S "EDINBURGH" ($17\frac{3}{4}'' \times 13\frac{1}{2}''$)
from *Civitates Orbis Terrarum* c. 1574

This list is not inclusive. There are many anonymous charts that have been omitted, and some of the notable ones relating to America which will be dealt with later. The British Museum is fortunately very rich in extant portolanos, about an eighth of the known total.

AUTHORITIES

AIRENTI (P. M. G.). *Osservazioni . . . sopra la Tavola Peutingeriana*, Roma, 1809.

ANDREWS (M. C.). "Study and Classification of Medieval Mappæ Mundi" (*Archælogia*, Vol. 75, 1925).

—— "Rathin Island in the Portolan Charts" (*J.R.S.A of Ireland*, Vol. XV, 1925).

BABCOCK (W. T.). *Legendary Islands of the North Atlantic, a study in mediæval geography*, N.Y., 1922.

BEAZLEY (C. R.). *Dawn of Modern Geography*, 3 vols., 1897–1906.

BEVAN (W. L.) and H. W. PHILPOTT. *Mediæval Geography: an Essay in Illustration of the Hereford Mappa Mundi*, 1873.

BLASQUEZ (A.). "Mapas Antiguos adquiridos por la Sociedad Bilbaina y un Mapa de Juan Olives de 1591" (*Bull. de la Real Soc. Geogr., Madrid*, 1918).

BRITISH MUSEUM. Four Maps of Great Britain designed by Matthew Paris, 1928.

BUNBURY (E. H.). *Ancient Geography*, 2 vols., 1879.

BURON (E.). *Ymago Mundi de Pierre d'Ailly*, 3 vols., Paris, 1930.

CARACI (G.). "The Cartographical Work of Grazioso Benincasa" (*R.G.S.J.*, 1936).

—— "Cimeli cartografica sconosciuta esistenti a Firenze carta nautica di Diego Homen" (*La Bibliophila*, 1928).

CRIVELLARI (G.). *Alguni cimeli della Cartografia Medievole existenti a Verona*.

CRONE (G. R.). *The Hereford World Map*, R.G.S., 1948.

D'AVEZAC (M.). *Atlas Hydrographique de 1511 du Génois Vesconte de Maggiolo extrait des Annales de Voyages, Juillet 1870*, Paris, 1871.

—— *Note sur un Atlas Hydrographique MS. exécuté à Venise dans le XV siècle conservé au Musée Britannique*, Paris, 1850.

—— *La Mappemonde du VIII siècle de Saint Béat de Liebana*, Paris, 1870.

—— *Note sur la Mappemonde historiée de la Cathédrale de Hereford*, Paris, 1862.

—— *Aperçus historiques sur la Rose des Vents*, Rome, 1874.

DESTOMBES (M.). *Monumenta Cartographica Vestustioris Aevi Mappemondes A.D. 1200–1500*. Amsterdam, Israel, 1964.

DOURADO (Fernao Vaz). Repro. of Atlas dated at Goa 1571, fol. Porto (1949).

FISCHER (T.). *Raccolta di Mappamundi e Carte Nautiche 13th–16th Cent.*, 1871–81.

—— *Sammlung mittelalterlicher Welt- und Seekarten italienischen Ursprungs und aus italienischen Bibliotheken und Archiven*, Venice, 1886.

GOUGH (R.). *British Topography*, 1780.

HAMY Collection of Portolan Charts of XV, XVI and XVII centuries, N.Y. Anders & Auction Co., 1912.

HANTSCH (V.) and L. SCHMIDT. *Kartographische Denkmäler König. Offent. Bibli. zu Dresden*, Leipzig, 1903.

JOMARD (E. F.). *Les Monuments de la Géographie*, Paris, 1842–62.

KAMMERER (A.). *La mappemonde Lopo Homen et l'Atlas Miller*, Geogr. Jnl., 1939.

KARSTENS (H.). *Die Ebstorfer Weltkarte*, Geogr. Anz., 1937.

KIMBLE (G. H. T.). *Geography in the Middle Ages*, 1938.

KRETSCHMER (K.). *Die italienischen Portolane des Mittelalters*, Berlin, 1909.

—— "Marino Sanudo der Ältere und die Karten des Petrus Vesconte" (*Zeitschrift der Gesellschaft für Erdkunde zu Berlin*, 1891).

—— "Die Atlanten des Battista Agnese" (*Zeitschrift der Gesellschaft für Erdkunde zu Berlin, Bd. XXI*, 1896).

—— "Handschriftliche Karten der Pariser National Bibliothek" (*Zeitschrift der Gesellschaft für Erdkunde zu Berlin*, 1911).

—— *Eine neue mittelalterliche Weltkarte der Vatikanischer Bibliothek*, Berlin, 1891.

—— *Die Katalanische Weltkarte der Bib. Estense ʒu Modena*, Berlin, 1897.

LA COSA (Juan de). Portolan World Chart, 1500. Reproduced Madrid, 1892.

LELEWEL (J.). *Géogr. du Moyen Age*, 5 vols. and atlas, Brussels, 1850–70.

—— *Pythéas de Marseilles et la géographie de son temps*, Brussels, 1836.

MAGNAGHI (A.). "Angellinus de Dalorto" (*Riv. Geogr. Ital.*, 1897).

—— "L'Atlante manoscritte di Battista Agnese nella Bib. R. di Torino" (*Riv. Geogr. Ital.*, 1908).

—— *Sulle origine del Portolano normale del Medio Evo*, Firenze, 1909.

MAGNOCAVALLO. "La Carta de Mari Mediterraneo di Marino Sanudo il Vecchio" (*Bull. della Soc. Geogr. Italiana*, 1902).

MALONE (Kemp). "King Alfred's North: a study in mediæval geography" (*Speculum, Vol. V.*, Cambridge, Mass., 1930).

MARCEL (G.). *Recueil de portulans*, Paris, 1886.

—— *Choix de cartes et de mappemondes XIV et XV siècles*, Paris, 1896.

MILLER (Konrad). *Mappaemundi: die ältesten Weltkarten*, 6 vols., Stuttgart, 1895–8.

—— *Die Weltkarte des Castorius*, Ravensburg, 1888.

—— *Mappae Arabicae*, Stuttgart, 1926–7.

MORI (A.). "La cartografia dell' Italia dal secolo XIV al XVIII" (*Boll. R. Soc. Geogr. Ital.*, 1930).

NORDENSKIÖLD (A. E.). *Facsimile Atlas*, Stockholm, 1889.

—— *Periplus*, Stockholm, 1897. Reprint Franklin, New York.

RONCIERE (C. de la). "L'Atlas Catalan de Charles V" (*Biblio. de l'Ecole de Chartres*, Tome LXIV, 1903).

ROSS (D.). "Arab map of the British Isles" (*Antiquity*, 1937).

ROXBURGHE CLUB. *Itineraries of William Wey to Jerusalem, 1458 and 1462*, 1867.

SANTAREM (Vicomte de). *Essai sur l'Histoire de la Cosmographie et de la Cartographie pendant le Moyen Age et . . . après les Découvertes du XV Siècle*, 3 vols., 8vo, 1848–52. *Atlas XI–XVII Siècle*, Paris, 1849–52.

STEVENSON (E. L.). *Portulan Charts, their origin and characteristics*, New York, 1911.

—— *Maps Selected to Represent the Development of Map-making from 1st to 17th Century*, New York, 1913.

UHDEN (R.). "Die Weltkarte des Martianus Capella" (*Peterm. Mitt.*, 1930).

—— "Die Antiken Grundlagen der mittelalterlichen Seekarten" (*Imago Mundi I*, 1935).

—— *Die Weltkarte des Isodorus von Seville Mnemosyne*, 1935.

—— "An unpublished Portolan chart of New World 1519" (*Geogr. Jnl.*, 1938).

UZIELLI (G.) and P. AMAT DI S. FILIPPO, *Studi Biografici e Bibliografici sulla storia della Geografia in Italia*, Roma, 1882.

VIDAGO (J.). *A Ilha do Brasil*, Horta, 1938.

WAGNER (H. R.). "The Manuscript Atlases of Battista Agnese" (*Bibliographical Soc. of America*, Vol. 25, 1931).

WESTROPP (T. J.). "Brasil and the Legendary Islands of the North Atlantic" (*Proc. Roy. Irish Acad.*, Vol. 30, 1912–13).

WIESER (F. R. von). *Die Weltkarte des Albertin de Virga*, Innsbruck, 1912.

WINSOR (J.). "Baptista Agnese and American Cartography in the XVI Cent." (*Proc. of Mass. Hist. Soc.*, 1897).

WINTER (H.). "On the Real and the Pseudo-Pilestrina Maps and other early Portuguese maps in Munich" (*Imago Mundi IV*, 1947).

—— "A Late Portolan Chart at Madrid and late Portolan charts in general" (*Imago Mundi VII*, 1950).

WRIGHT (J. K.). *Geographical Lore of the Time of the Crusades*, 1925.

ZURLA (Cardinal Placido). *Il mappamondo di Fra Mauro Camoldese*, Venice, 1806.

15   G. M. CONTARINI'S MAP OF THE WORLD (24¾″ × 16½″)
The earliest printed map to show America, 1506. (*B.M. Maps C2 cc*)

16 CAMOCIO'S MAP OF MALTA (6¾″ × 10″)

CHAPTER III

Italy

AS in most parts of the arts and sciences, Italy's contribution to map-making has been immense, both on the theoretical and practical side. Various causes combined to make Italy the centre of the map-making industry at an early date: her favourable geographical position in the centre of the civilised world, the skill and daring of her navigators and explorers, and finally the tradition of craftsmanship of her artisans. For example, the first substantial addition to a knowledge of the Far East was due to the Polos, and practically all the east coast of America was discovered by Italians—Columbus, Vespucci, John Cabot and Verrazano, respectively Genoese, Florentine and Venetian. Italian bankers and merchants traded over most of Europe, and had footholds in Asia, and the wealth and information they gathered flowed back to Rome, Venice, Genoa and other cities.

No appreciation of Italian effort in the geographical field is complete without reference to manuscript records. Two of the most famous of these are the world map of Marino Sanudo, 1306, and the large world map of Fra Mauro, of 1457–9, executed in the convent of Murano, near Venice. Over 6 feet in diameter, it is crammed with detail, in form resembling the mediæval mappæmundi, but greatly amended for Mediterranean and western Europe, and showing the new discoveries of the Portuguese. But the most important part of this manuscript material is the sea charts, known under the name of portolan charts or portolans. These portolans, at first confined to the Mediterranean basin, were gradually extended to include western Europe, then the discoveries in Africa and America, finally becoming hydrographical charts of the world. About 500 examples are known to survive at present, of which number no less than 52 are preserved in the British Museum. A brief list of these chart-makers is given in Chapter II.

To turn to printed maps, to Italy belongs the honour of being the first to revive an interest in classic geography, the greatest extant work of the ancients, the *Geographia* of Ptolemy, being first printed in Italy, Bologna 1477, Rome 1478, Florence 1482 and Rome 1490. The maps in these editions, particularly the Rome editions, are beautiful examples of copper-plate engraving, superb testimonials of Italian craftsmanship without the picturesque but unscientific monsters of the mediæval maps or the addition of adventitious decoration of later work, relying for their beauty solely on the delicacy of their execution and the fineness of the material employed. Eleven further editions were printed in Italy at intervals.

Another popular form of atlas was that dealing with islands, e.g. those of Bartolomeo dalli Sonetti, the *Isolario* of Benedetto Bordone, published in Venice 1547, with 3 folding and 107 maps in the text; Camocio's *Isole famose* with 87 maps, and Tomaso Porcacchi's *l'Isole più famose del Mondo*, Venice 1572, reissued 1590 and later. Another excellent work was the

Geografia of Livio Sanuto, also a Venetian production. Printed in 1588, this was the most important publication on Africa issued up to this date.

Lafreri, Forlani and Bertelli

The middle decades of the 16th century witnessed great activity in the production and publication of separate maps of all parts of the world, the main centres of the industry being Rome and Venice, the former tending to be more local and topographical, the latter more cosmopolitan. So many cartographers, editors, engravers and publishers were at work then that it is only possible in a short account to mention a few: Bell' Amato of Siena, F. and D. Bertelli of Venice, Bordone of Padua, Seb. Cabot, Contarini and Camocio (16) of Venice, Cartaro of Rome, Cimerlino of Verona, Coppo, a Venetian, De Musis, Duchetti, and Floriani of Udine, Forlani of Verona, Gastaldi, a Piedmontese, Giovio of Como, Giolito, Guicciardini, Porro, Lafreri of Rome, Pirro Ligorio of Naples, Luchini, Marcollino, Moletti, Nelli, Ruscelli, Salamanca, Sylvanus, Tramezini, Valgrisi, Zaltieri and Zenoi.

17 *City Plan from Bordone's* Isolario, 1547

The most outstanding of these was Giacomo Gastaldi. Born at Villa Franca in Piedmont, he settled in Venice, and was appointed cosmographer to the Republic. No less than 109 separate map publications due to his industry have been traced so far. His more important works include maps of the world, maps to an edition of Ptolemy's *Geography*, and large maps of Africa, Asia, Europe, Italy, Poland and Lombardy. (18)

Fortunately for posterity, some unknown individual conceived the idea of accumulating a certain number of these loose maps and binding them in atlas form (folding the large maps to size), thus preserving many maps that would otherwise have been destroyed. Many such collections still exist, between 60 and 70 having so far been traced. These collections were formed independently, no two collections being alike either numerically or in the maps selected. They have one common feature, the maps being arranged in the old Ptolemaic system of precedence. They are in fact the first modern atlases, slightly antedating the *Theatrum* of Ortelius. About 1570–2 a finely engraved title-page was issued—"Tavole Moderne di Geografia de la maggior parte del mondo di diversi autori . . . con studio et diligenza in Roma." The fact that this title bore a Rome imprint (though no publisher is given) and that Lafreri of Rome printed a list of maps on sale at his establishment that agrees more or less with the contents of such collections, led to the rather loose appellation of Lafreri Atlases to such volumes. This is far from correct, Forlani, Bertelli and Duchetti at least having a claim in this respect.

In spite of the number of known examples, these volumes of 16th-century Italian maps very rarely occur for sale, though it is possible to acquire odd maps of the period occasionally. They are always worth possessing, adding a distinction to any collection. They are mostly fine examples of copper-plate engraving, in the severe but elegant Italian style without ornamentation, the calligraphy of great beauty. Intrinsically their value varies: some, naturally the Italian maps, were based on original surveys; but for countries north and west of the Alps the Italian map-makers used or reproduced the work of foreign geographers, sometimes in full, sometimes on a reduced scale, and quite a number of these early cisalpine maps are only known from their Italian reductions. Gastaldi is in a class by himself, both his maps and his geographical theories being issued in whole and in part by cartographers of other lands.

In the midst of this seeming prosperity various influences were slowly but surely undermining Italian supremacy. The gradual but increasing transfer of the trade routes of the world from the Mediterranean to the Atlantic seaboard, and the consequent elimination of Italy as a distributive centre, destroyed at the same time her sources of wealth and information. The year 1570 marks the visible turning-point as regards map production, for in May of that year Ortelius produced the first edition of his celebrated atlas in Antwerp. Ortelius was closely followed by Mercator (24) and Hondius, and pre-eminence in map production passed from Italy to the Low Countries. From this point the decline was rapid.

Later Italian work, though often excellent in execution, was mainly imitative, apart from its contribution to local geography. In the 17th century, Giovanni Antonio Magini's *Italia*, Bononiae, 1620, is worthy of note. It is in a somewhat florid style, with an engraved title and 60 maps. Ten of these were engraved by Benjamin Wright, an Englishman. But the most important and rarest of the 17th-century atlases is the *Arcano del Mare*, of Sir Robert Dudley. Dudley, an illegitimate son of the Earl of Leicester and brother-in-law of Cavendish the circumnavigator, an excellent mathematician and navigator himself, turned Catholic and settled in Florence. His great work, the *Arcano del Mare*, first appeared in 1646–7, a second corrected edition appearing in 1661. Beautifully engraved by Antonio Francesco Lucini, who spent eight years in engraving the copper-plates, it is without ornamentation, and far in advance geographically of other atlases of the period. Towards the end of the century Giacomo Rossi published the *Mercurio Geografico* (Rome, 1692) in two folio volumes with 155 maps, and V. M. Coronelli issued his atlas *Atlante Veneto*, 4 volumes, folio, Venice, 1690–6, with over 300 maps. Besides sea charts, it contains several large plates of the ships of various nations, portraits of the Doges and other notabilities and views of Venice. He likewise issued an *Epitome Cosmografica* in quarto, 1693, with 38 maps and plates.

In the 18th century an atlas of note was P. Santini's *Atlas Universal*, in two large folio volumes 1776, and Antonio Zatta's *Atlante novissimo*, Venice 1779–85, in quarto, with 218 maps. The most spectacular from a technical point of view was, however, Rizzi Zannoni's *Atlante Marittimo delle due Sicile*, 1793. Rizzi Zannoni's work is remarkable for a marvellously minute attention to detail, and amply deserves the commendation passed upon it by Sir George Fordham.

AUTHORITIES

ALMAGIÀ (R.). "Primo saggio storico di cartografia abruzzese" (*Rivista abruzzese di Sc. Lettere e Arti*, 1912).
—— *Studi storici di cartografia napoletana*, 2 vols., Napoli, 1913.
—— *La Carta d'Italia di G. A. Vavassori*, Firenze, 1914.
—— *La cartografia dell' Italia nel Cinquecento*, Firenze, 1915.
—— *Una Carta della Toscana della metà del secolo XV*, Firenze, 1921.
—— *Monumenta Italiae Cartographica*, Firenze, 1929.
——— "Intorno ad una raccolta di carte cinquecentesche di proprietà del Lloyd Triestino" (*L'Universo*, 1927).
——— "Due carte nautiche manoscritte nella Bib. di Cefalu" (*Riv. Geogr. Ital.*, 1937).
—— "Intorno al un grande mappamondo perduto di Giacomo Gastaldi (1561)" (*La Bibliofilia*, 1939).
—— *Alcune stampe geografiche italiane dei secoli XVI e XVII oggi perdute*, Milano, 1940.
——— *I mappamondi di Enrico Martello e alcuni concetti geografici di Cristoforo Colombo*, Firenze, 1941.
—— *Un prezioso cimelio della cartografia italiana: il planisfero di Urbano Monti*, Firenze, 1941.

ALMAGIÀ (R.). *L'Opera geografica di Luca Holstenio*, Vaticano, 1942.

—— *Monumenta Cartographica Vaticana*, Vols. I–III, Bib. Apostolica Vaticana, 1944–52.

—— *Osservazioni sull' opera geografica di Francesco Berlinghieri*, 1945.

—— "Quelques questions au sujet des cartes nautiques et des portulans d'après les recherches récentes" (*Arch. Intern. d'Histoires des Sciences*, 1948).

—— "The First 'Modern' Map of Spain" (*Imago Mundi V*, Stockholm, 1948).

—— "The Atlas of Pietro Coppo, 1520" (*Imago Mundi VII*, 1950).

—— *Vicenzo Coronelli: discorso ufficiale in Palazzo Ducale il 21 Maggio 1950*, Venezia, 1951.

—— *Note sulla cartografia dell' Italia nei secoli XV e XVI*, Roma, 1951.

ARMAO (Ermano). *Vincenzo Coronelli: cenni sull'uomo e la sua vita*, Firenze, 1944.

BAGROW (L.). *A. Ortelii Catalogus Cartographorum*, 1928–30.

—— "Maps from Home Archives of Descendants of a friend of Marco Polo" (*Imago Mundi V*, 1948).

BEANS (G. H.). *Fragments from a Venetian Collection of Maps*, 1931.

—— *A Large World Map dated 1569 by J. F. Camotius*, 1933.

—— *Maps ex Duke of Gotha Collection*, 1935.

—— *Some 16th-century Watermarks*, 1938.

BERTOLINI (G. L.). "Note sulla carta del Territorio Trevigiano nell'Atlante Magini" (*Boll. Soc. Geogr. Ital.*, 1906).

BEVILACQUE (G.). *Cenni storici su di alcuni geo-idrografi Anconitani*, Ancona, 1862.

BLESSICH (Aldo). *Un geografo italiano nel sec XVIII: Giov. Anton. Rizzi Zannoni (1736–1814)*, Roma, 1898.

BRITISH MUSEUM. *A Map of the World, designed by G. M. Contarini, engraved by F. Roselli*, 1924. Second edition, revised, 1926.

BRUZZO (G.). "Di Grazioso Benincasa e del suo Portolano, Firenze" (*Riv. Geogr. Italiana*, 1897).

BUONARRO (G.). *I due rarissimi globi di Mercatore nella Biblioteca di Cremona*, Notizia Cremona, 1890.

CARACI (G.). *Tabulae Geographicae Vestustiores in Italia adservatae*, 1926–8.

—— "An Unknown Nautical Chart of Grazioso Benincasa 1468" (*Imago Mundi VII*, 1950).

CASTELLANI (C.). *Catalogo Ragionato delle più rare o più importanti opera geographice a stampa che si conserva nella Biblioteca del Collegio Romano*, 1876.

COLOMBO (Cristoforo) *e la scuola cartografica genovese (Consiglio Nazionale delle richerche) Magrini, Picotti, Revelli, Issel, Grosso, Marengo*. 3 vols., Genova, 1937.

CRINÒ (S.). "La prima carta corografica inedita del Giappone portata in Italia nel 1585" (*Riv. Marit.*, 1931).

—— *I Planisferi di Francesco Rosselli dell' epoca delle grandi scoperte geografiche*, Firenze, 1940.

CRIVELLARI (G.). *Alcuni Cimeli della Cartografia Medievale esistenti a Verona*, Firenze, 1903.

D'AVEZAC. *Atlas Hydrographique de 1511 du Génois Vesconte de Maggiolo*, Paris, 1871.

DAWSON (S. C.). *Memorandum upon the Cabot Map (World 1544)*, Ottawa, 1898.

DEL BADIA (J.). *Egnazio Danti cosmografo e matematico e le sue opere in Firenze*, Firenze, 1898.

DONATI (RAMBERTO). "Natale Bonifacio" (*Arch. Storica per le Dalmazia*, 1927).

—— "Un libro sconosciuto illustrato da Natale Bonifacio sibenicense" (*Arch. Storica per le Dalmazia*, 1930).

DURAZZO (P.). *Il Planisfero di Giovanni Leardo*, Mantova, 1885.

EDWARDS (Francis). *Description of a Recently Discovered Lafreri Atlas* (1933).

EHRLE (F.). *Memoirs on Lafreri's Plan of Rome, 1577*, Rome, 1908.

EMILIANI (M.). "Le carte nautiche dei Benincasa cartografi anconitani" (*Boll. R. Soc. Geogr. Ital.*, 1936).

FEROSO (L.). *Grazioso Benincasa marinaro e cartografo Anconitano del secolo XV*, Ancona, 1884.

FIORINI (Matteo). *Sfere Terrestre e celesti di Autore Italiano oppure fatte o conservati in Italia*. Roma, 1899.

GALLO (R.). "Antonio Florian and his Mappemonde" (*Imago Mundi VI*, 1949).

—— "Giovan. Francesco Camocio and his Large Map of Europe" (*Imago Mundi VII*, 1950).

GORI (P.). *Paolo del Pozzo Toscanelli (1397-8–1482*, Firenze, 1898).

GRANDE (S.). *Notizie sulla Vita e sulla Opera di Giacomo Gastaldi*, Torino, 1902.

—— *Le Carte d'America di Giacomo Gastaldo*, Torino, 1905.

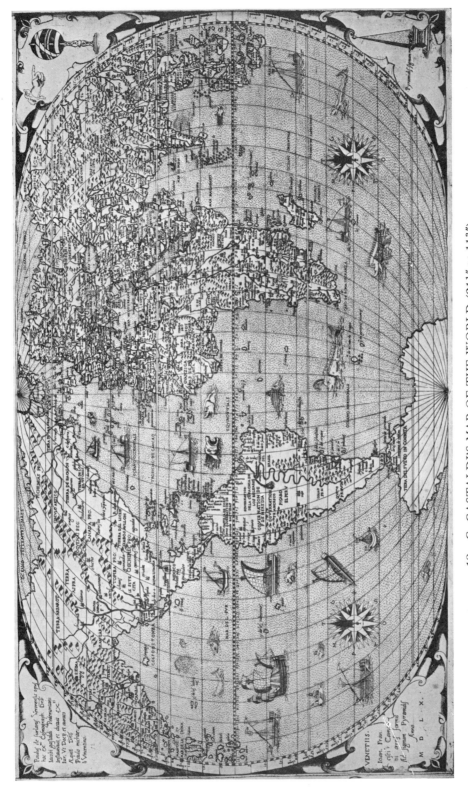

18 G. GASTALDI'S MAP OF THE WORLD ($21\frac{1}{4}'' \times 11\frac{3}{4}''$)

Venice, Camocio, 1560. (*B.M. Maps C7 e1(2)*)

19 P. DE LA HOUUE'S ITALY (21½″ × 15¼″). (Plancius, 1620)

GRIBANDI (P.). *Inventario dei MSS. geografici della Biblioteca Palatina di Parma*, Parma, 1907.

HEAWOOD (E.). "An Undescribed Lafreri Atlas and Contemporary Venetian Collections." (*R.G.S.*, Vol. 73, 1929).

—— "Another Lafreri Atlas" (*R.G.S.*, Vol. 80, 1932).

—— "Florentine World Maps of Francesco Roselli" (*Geog. Jnl.*, 1940).

HUELSEN (C.). *Saggio di Bibliografia ragioneta delle piante iconografiche e prospettille di Roma*, 1551–1748, Firenze, 1933.

LYNAM (E.). *The Map of the British Isles of 1546*, Jenkintown, 1934.

—— "Gianpietro Contarini's map of Europe and Asia, 1564," (*B.M. Quarterly*, 1938).

—— *The First Engraved Atlas of the World, the Cosmographia of Claudius Ptolemaeus Bologna, 1477*, Jenkintown, 1941.

MAGNAGHI (A.). *La Carta Nautica costruita nel 1325 da Angelino Dalorto*, Firenze, 1898.

—— *Contributi alla storia della cartografia d'Italia sulle origini del portolano normale nel medio evo e della cartografia dell' Europa occidentale*, Firenze, 1909.

MAJOR (R. H.). *On a Mappemonde by Leonardo da Vinci*, 1865.

—— *Voyages of the Venetian Brothers Nicolo and Antonio Zeno*, 1873.

MANNO (A.) and V. PROMIS. *Notizie di Jacopo Gastaldi*, Torino, 1881.

MARINELLI (G.). *Saggio di cartografia della regione Veneta*. Venezia, 1881.

MORI (A.). *Come Progredi la conoscenza Geografica della Toscana nel secolo XIX*, Firenze, 1899.

—— *Studi, trattative e proposte per la costruzione di una carta geografica della Toscana nella seconda meta del secolo XVIII*, Firenze, 1905.

—— *La cartografia ufficiale in Italia e l'Istituto geografico militare. Notizie storiche raccolte e ordinate*, Roma, 1922.

—— "Le carte della Toscano di D. Stefano Buonsignori" (*La Bibliofila*, 1907).

MOTZO (B. R.). *Il Compasso da navigare, opera ital. della meta del Secolo XIII*, Cagliari, 1947.

NUNN (G. E.). *Antonio Salamanca's version of Mercator's Map of 1538*, Jenkintown, 1935.

—— *World Map of Francesco Roselli*, 1928.

PARMA. *Inventario dei manoscritti geografici della R. Bib. Palatina di Parma*, 1907.

PELLATI (N.). *Contribuzione alla storia della cartografia geologica in Italia*, Roma, 1904.

PETRUCCI (F. Bargagli). *Il Mappamondo di Ambrogio Lorenzetti per il Palazzo Pubblico di Siena*, Siena, 1914.

PEZZA (F.). *Profilo Geografico della Bulgaria Italiana nel alto Medioevo*, Novara, 1936.

PORENA (F.). *Schiarimenti intorno al passagio del primato cartografico dall' Italia ai Paesi Bassi nel secolo XVI*, Napoli, 1905.

PULLE (L.). *Mappamondo Catalano della Estense*, Modena.

QUARITCH (Bernard, Ltd.). *The Speculum Romanae Magnificentiae of Antonio Lafreri, together with a Description of a Bertelli Collection of Maps* (1926).

ROLAND (Dr. F.). "Antoine Lafrery (1512–1577)" (*Notice Historique Besançon*, 1911).

ROSSI (G. B. de). *Sopra el cosmografo Ravennate e gli antichi geografi citati da lui*, Roma, 1852.

SHIRAS (W.). "The Yale 'Lafrery Atlas' " (*Yale Univ. Lib. Gazette*, 1935).

SORANZO (G.). *Bibliografia veneziana*, Venezia, 1885.

STEVENSON (E. L.). *Portolan Charts*, New York, 1911.

TOOLEY (R. V.). "Maps in Italian Atlases of the XVI Century" (*Imago Mundi III*, 1939).

UZIELLI (G.) and P. AMAT DI S. FILIPPO. *Studi Biografici e Bibliografici sulla della Geografie in Italia*, Rome, 1882.

VIGNAUD (II.). *Bibliografia della polemica concernente Paolo Toscanelli e Cristoforo Colombo . . . Trad. con intro. e aggiunte da G. Uzielli*, Napoli, 1905.

ZURLA (Don Placido). *Il Mappa Mondo da Fra Mauro*, Venice, 1806.

—— *Dissertazione intorno di Viaggi e Scoperte di Nicolo e Antonio Zeni*, Venezia, 1808.

—— *Sulle Antiche Mappe idro-geografiche lavorato in Vinezi*, Venezia, 1818.

CHAPTER IV

Germany, Austria and Switzerland

GERMANY holds a peculiar position in the history of cartography. At no period has she visibly dominated map production as Italy, the Low Countries, France and England have each in their turn. Yet her contribution to the advance of geographical science has been both extensive and valuable. That this is not so immediately apparent is due to the general absence of atlases in the early period (for, apart from editions of Ptolemy's *Geography*, German production was limited in the main to globes and separate map publications) and to the fact that the earliest Germanic wood-cut maps are amongst the rarest of cartographical monuments, and were adopted and popularised in other countries. Lastly, many of her great cartographers worked abroad, and their work has become associated to a certain extent with their country of domicile.

The first outstanding landmark in German cartography was the printing of Ptolemy's *Geography* at Ulm, in 1482. It is notable that this, the first production in Germany, was an improvement on the Italian original. To the 27 maps of the ancient world as conceived by Ptolemy, the editor Nicolaus Germanus added 5 new maps based on contemporary knowledge, besides revising the original maps before they were engraved. A second edition appeared in 1486, also in Ulm. An even more ambitious edition was issued in 1513 in Strassburg. With 20 additional maps drawn according to the new knowledge, this may be said to be the first atlas compiled from contemporary sources since the age of the classics, and in this sense is the first modern atlas. It is the most remarkable of the ancient editions of Ptolemy from a geographical point of view, giving to the public the best picture of the world possible at that date. The world map is based on Portuguese sources, and the Admiral's chart is the first separate representation of the American continent in an atlas. Re-issues followed in 1520, 1522 and 1525, all at Strassburg, the last two editions with additional information.

Thus at an early date German cartographers were working on original material, and this brings us to the more important part of their work, the separate publications. The most venerable figure among these early workers was the Cardinal Nicolas Cusanus, whose birth name was Nicholas Cryftz. He was born in 1401 in Cues, a little town on the Moselle. Educated at Heidelburg and Padua, he studied mathematics and astronomy, and became one of the most learned men of his day. He travelled extensively, his itineraries including Rome, Paris, Prague, Cologne, Basle and Constantinople. In 1448 he was elevated to the rank of Cardinal. He died in 1464 and left his manuscripts to his native town of Cues. One of these, a MS. map, was printed posthumously in 1491 and was the first modern map of Germany. (20)

The end of the century saw the production of two important works—a map of the world by Henricus Martellus Germanus (undated), and a globe by Martin Behaim, 1492.

Martin Behaim, a native of Nürnberg, was born in 1459. In 1484 he proceeded to Portugal, at that time the main centre of geographical information. Here he married, and later settled in the Azores. In 1490 he returned to his native city of Nürnberg and resided there till 1493. During his visit he was commissioned by the Town Council to construct a terrestrial globe. This he completed in 1492. This globe has achieved fame as the last notable representation of the world prior to the discovery of America. As the work of Ptolemy summed up the knowledge of the Greeks, so the globe of Behaim showed the extent of available geographical knowledge at the close of the 15th century. It marked the end of one age and the beginning of a new, this new age being represented by a better delineation of the coast-lines of Europe, based on the portolani, and the charting of the southern coasts of Africa as a result of the voyage of Bartholomew Diaz round the Cape in 1486. Behaim returned to Portugal in 1493 and there he died in 1507.

The 16th century was one of the most brilliant periods in German cartography, due in part to the skill of such men as Peuerbach, Apian and Stöffler in devising improved surveying instruments. Research in recent years has revealed the existence of many previously unsuspected source maps. In fact, so prolific was German output during this period that it is only possible in this brief sketch to mention some of the more important scholastic centres and a few of their main productions.

Nürnberg (or Nuremberg), one of the great cultural centres of Germany during the Middle Ages and the Renaissance, was especially prominent both in the theory and practice of geographical science. Her artisans excelled as goldsmiths and metal workers, and produced the best globes and the finest compasses in Europe. Her scholars and geographical practitioners included Regiomontanus, Behaim, Etzlaub, Hirschvogel, Schöner, Schedel and Zundt. It was at Nürnberg that a new type of map, the road map, was developed. About the year 1500 Erhard Etzlaub, compass maker, compiled a map of Central Europe entitled "Der Rom Weg." This map showed the main travel routes to Rome by means of points from town to town, mile by mile through the whole of Germany from Denmark in the north to Naples in the south, with three passes over the Alps, the Semmering, the Brenner and the Splugen. From east to west it embraced the country between Paris and Cracow. To modern eyes these two cities are apparently reversed at first glance, as Rome is placed towards the top of the map and the north at the foot. Etzlaub also issued a road map of the neighbourhood of Nürnberg, and this, like "Der Rom Weg," was engraved on wood. These two maps are said to be the oldest printed maps of their kind. It is certainly curious that this important feature of map-making, the road or route, was not again inserted in printed maps till a considerably later period.

Johann Schöner was not a native of Nürnberg, being born at Carlstadt, in Lower Franconia, in 1477 and educated at Erfurt. He settled in Nürnberg in 1504, and there he studied mathematics and astronomy. The friend of Copernicus and Melanchthon, he started his career as a Catholic priest, but being discharged from his ecclesiastical duties for too great an attention to mathematical pursuits, he turned Protestant, was appointed professor of mathematics at Nürnberg University, and died in 1547. During his life he compiled no less than 46 publications on astronomy, astrology, geography, mathematics and medicine. His chief claim to fame to-day rests on the four globes he constructed in 1515, 1520, 1523 and 1533. It has been stated that the 1515 globe indicates a knowledge of a passage round South America prior to that of Magellan (who first circumnavigated the globe in 1520). The far more simple and natural assumption is that it indicated a belief or hope rather than an established fact at that time, for Schöner combined theory with fact in his various works. This is borne out by the fact that he also indicates a North-West Passage, a belief held from earliest times, but not proved till centuries later.

Another flourishing school of cartography was centred in Cologne, the works of Braun and Hogenberg, Quad and Vopel adding a lustre to its fame. Kaspar Vopel, born in 1511 in Medebach in Westphalia, died in 1564 in Cologne. In 1542 he constructed a fine globe map of the world printed in twelve segments. In 1555 he issued a large map of the Rhine on five wood-cut leaves, and in the same year a large-scale map of Europe which is known only from a re-issue on nine woodcut leaves published in 1566.

M. Quad produced a small oblong atlas *Europae totius orbis Terrarum* in 1592 (*Geographisch Handtbuch*, 1600 (82 maps)), and *Fasiculus geographicus* in 1608. But the most celebrated production of the Cologne geographers was the *Civitates Orbis Terrarum* of Braun and Hogenberg in five folio volumes, 1573–98. This forms a wonderful picture book of Europe in the 16th century. It was the first general collection devoted solely to topographical views. The plates were engraved by Hogenberg and others. Principally concerned with the towns of Europe (14), a few in Asia, Africa and even America were included. Britain was represented by London, Edinburgh, Canterbury, Oxford, York, Chester, Exeter etc. For many of the towns depicted the *Civitates* gave the first engraved view. Apart from its main topographical value, this magnificent work is of great interest as a record of the domestic life of the period, the town views, heraldic coats-of-arms, rural scenes, land and water transportation, public buildings, etc. It can make an equal appeal to the historian, geographer, the student of costume or the plain admirer of decorative art. A supplementary sixth volume was published in 1618. Copies were issued both plain and coloured, the latter being the most valuable, though not necessarily the most desirable.

Martin Waldseemüller, born about 1470 at Radolfzell on the Bodensee, studied at Freiburg, where he matriculated in 1490, and finally settled at St. Dié in Lorraine, at that time a centre of learning under the protection of Duke Rene II. Waldseemüller has achieved romantic fame as the godfather of America, for it was he who first suggested, in his *Cosmographio Introductio, 1507,* that the name America should be applied to the New World, after Amerigo Vespucci. Waldseemüller's more solid achievement was the compilation of a large map of the world in 1507. This was engraved on twelve wood-blocks, with the title "Universalis Cosmographica." Of 1,000 copies printed, only one is known to survive at the present time. This map was both popular and influential, being extensively copied on a smaller scale, and imitated by Stobnicza, Apian, Schöner, Reisch, Grynaeus, Münster, Gemma Frisius and others. In the same year, namely 1507, Waldseemüller executed a small globe, and in 1511 a large-scale map of Europe, of which not a single copy is now known to exist. Waldseemüller was also responsible for the "modern" maps in the famous 1513 Strassburg edition of Ptolemy's *Geography.* His last important work was another world map known as the "Carta Marina." This was issued in 1516 on twelve woodcut sheets.

Another cartographer of note and innovator on an extensive but smaller scale was Sebastian Münster. Born in Hessen in 1489, he studied first in Heidelburg and later in Tübingen, and finally settled in Basle. His best-known works were his *Typus Cosmograph. Universalis,* 1532, 1537 and 1555, his *Geographia Universalis,* 1540, of which several editions were published, and his *Cosmographia,* which first appeared in 1544 and subsequently in other editions. Münster's *Cosmography* was illustrated with double-page folding woodcut maps and views of famous towns, and an endless number of smaller woodcuts enshrined in the text, depicting portraits of celebrities, costumes, plants, animals, etc. Many of these smaller woodcuts were purely figurative, the same woodcut being utilised several times in the course of the work. Münster was the first to introduce a separate map for each of the four then known continents, viz. Europe, Asia, Africa and America (80), a practice that has been almost universally followed.

The output of Münster was considerable, as apart from the labour involved in the preceding works, he was responsible for editions of Solinus and Pomponius Mela illustrated with maps, besides separate publications, such as a map of Germany in 1525, a world map in 1532 and a map of Europe in 1536.

20 CARDINAL CUSA'S CENTRAL EUROPE, 1491 (22″ × 16½″)

(*B.M. Maps C2 a1*)

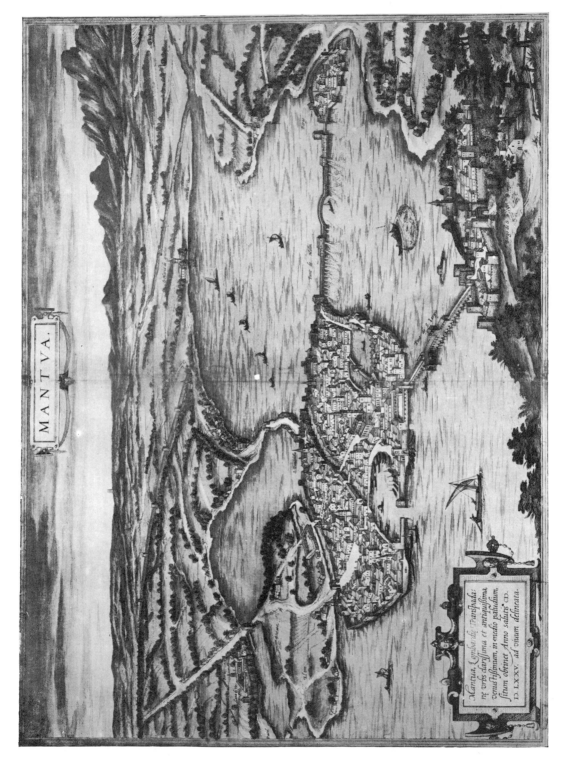

MANTVA.

21 BRAUN AND HOGENBERG'S PLAN OF MANTUA (19½″ × 14″)

from *Civitates Orbis Terrarum*; 1575

Two further cartographers of importance during the same era were Peter and Philip Apian. Peter Apian was born in 1495 at Leisnig in Saxony, studied astronomy and mathematics at Leipzig University, proceeded to Vienna, and finally settled in Ingoldstadt. His geographical works include maps of the world in 1520 and 1530 (the only known example of the latter being in the British Museum), a four-leaved woodcut map of Hungary in 1528 (of which likewise only one copy is known) and a map of Europe in 1534, of which no copy now survives.

Philip Apian, his son, achieved fame as the author of a fine large-scale map of Bavaria in 1563. This was copied at a much later date, namely, in 1761, on 40 leaves, and again in 1771 on 30 leaves. Apian himself made a reduction of his own map in 1568 on 24 leaves, and in 1576 he constructed a globe.

Other German geographers of the 16th century, whose work can be no more than mentioned, include Carl Heydans (map of Germany on 16 leaves, 1585), Martin Helweg (map of Silesia, 1561), Godfried Maschop (Munster and Osnaburg, 1568, on 9 leaves, used in reduced form by Ortelius), Johannes Mellinger (Thuringia 1568, Mansfeld 1571, and Luneburg 1593, used by Ortelius, de Jode and Mercator), Kaspar Hennebergen (Prussia, 1576), Ziegler (Palestine, 1532), Zoll (Europe) and Mathes Zundt (maps of Malta, Hungary, Cyprus, Corsica and various town plans).

Of the preceding maps of the 16th century, many on a large scale, some are known only by repute, others by a single, and few by more than three or four extant copies. To the collector they are unattainable, and even to the student they are not always available except in their place of custody, for few have been reproduced. The atlases, though rare, do occasionally occur for sale, and copies of the Ulm and Strassburg Ptolemies, Braun and Hogenberg's *Civitates* and Münster's *Cosmography* may be obtained, at a price.

In the 17th century Germany produced little of note in the cartographical sense, being submerged, like the rest of Europe, by the immensely competent, resourceful flood from the presses of Holland, and for the best maps of Germany or her provinces during this period it is necessary to turn first to the productions of Mercator and Hondius, and then to those of Blaeu and Jansson for the areas desired. German cartography did not, of course, suffer a total eclipse, but her output could not be compared with that of the previous century. One fine production was the atlas of Schleswig and Holstein, issued by Danckwerth and Mejer in 1652 with 40 maps. Johannes Mejer was a Danish mathematician and Royal Cartographer to Christian IV and Frederick III. Another interesting production was the atlas of the Archbishopric of Mainz, issued by N. Person in 1694.

The 18th century witnessed a considerable revival of activity: Anich and Hueber issued an *Atlas Tyrolensis* in Vienna in 1774 with 40 maps; M. Koops published a fine map of the Rhine on ten sheets in 1796; I. Müller a map of Hungary on twelve sheets in 1769, and Weigel a miniature atlas at Nürnberg in 1720. The most important and prolific map-makers in Germany in the 18th century were the Homann family (1702–1813). The founder and principal member was Johann Baptist Homann. He set up his headquarters in Nürnberg and quickly dominated the German market. Nor did he confine his efforts to his homeland, but produced general atlases covering the whole world. The Homanns produced a *Neuer Atlas* in 1714, a *Grosser Atlas* in 1737, and an *Atlas Maior*, with over 300 maps, about 1780. They likewise issued a special *Atlas of Germany*, with full-size plans of the principal cities, school atlases, and an *Atlas of Silesia* in 1750 with 20 maps. The maps of the Homanns were issued plain and coloured; the colouring when used was somewhat harsh and crude, the cartouches with which the maps were decorated being usually left plain. A few examples, no doubt for special clients, had the cartouches fully painted, and such examples were more carefully coloured and are very fine. The principal rival of the Homanns was Matthias Seutter, who published his *Atlas Novus* in Vienna in 1730, a *General Atlas* in Augsburg about 1735, and an *Atlas*

Minor in 1744. His maps are very similar both in execution and colouring to those of Homann.

Finally, it is to Germany that we owe some of the most important and extensive research work that has been done in recent years in the study of ancient cartography.

AUTHORITIES

ALPINE. *Essai d'Histoire de la Cartographie Alpine pendant les XV-XVIII siècles*, Grenoble, 1903.

AVERDUNK (H.) and H. MÜLLER-REINHARD. *Gerhard Mercator und die Geographen unter seinen Nachkommen*, Gotha, 1914.

BACKMANN (F.). *Die alten Städtebilder-Schedel bis Merian*, Leipzig, 1939.

BAGROW (L.). *A. Ortelii Catalogus Cartographorum*, 1928–30.

—— "Rust's and Sporer's World Maps" (*Imago Mundi VII*, 1950).

BURMEISTER (K. H.). *Sebastian Münster. Eine Bibliographie*, Pressler, Wiesbaden, 1964.

ENGELMANN (W.). *Bibliotheca Geographica*, Leipzig, 1858.

FISCHER (J.) and F. R. v. WIESER. *Die älteste Karte mit dem Namen America aus dem Jahre 1507 von M. Waldseemüller (Ilacomilus)*, Innsbruck, 1903.

GALLOIS (L.). *Les Géographes allemands de la Renaissance*, Paris, 1890.

GHILLANY (Dr. F. W.). *Geschichte des Seefahrers Ritter Martin Behaim*, 1853.

GRENACHER (F.). *Der Basler Druck der Cusa-Germania Karte*, Basler Volksblatt, 1942.

—— "Die Schweiz auf alten Karten" (*National Zeitung*, Basel, 1945).

GROB (Dr. R.). *Geschichte der schweizerischen Kartographie*, Bern, 1941.

GUNTHER (S.). *Erd- und Himmelsgloben, ihre Geschichte und Konstruktion*, Leipzig, 1895.

HANTSCH (V.). *Sebastian Münster. Leben, Werk, wissenschaftliche Bedeutung*, Leipzig, 1898.

—— *Die Landkartenbestände der K. O. Bib. zu Dresden*, Leipzig, 1904.

HEAWOOD (E.). "Glareanus, His Geography and Maps" (*Geogr. Jnl.*, 1905).

HERRMANN (A.). *Die ältesten Karten von Deutschland bis Gerhard Mercator*, Leipzig, 1940.

HORN (W.). "Sebastian Münster's Map of Prussia and the variants of it" (*Imago Mundi VII*, 1950).

IMHOFF (E.). *Die ältesten Schweizerkarten*, Zürich, 1939.

—— *Hans Konrad Gygers Karte des Kantons*, Zürich Atlantis, 1944.

ISCHER (T.). *Die ältesten Karten der Eidgenossenschaft*, Bern, 1945.

KRÜGER (H.). "Die Romweg-Karte Erhard Etzlaubs, 1492" (*Peterm. Mitt.*, 1942).

—— *Deutschlands älteste Strassenkarten Erhard Etzlaubs 15 u 16 Jhr.*, Berlin, 1942.

—— "Georg Erlinger von Augsburg als Kopist Etzlaubscher Strassenkarten" (*Peterm. Mitt.*, 1942).

LANG (W.). "The Augsburg Travel Guide of 1563 and the Erlinger Road Map of 1524" (*Imago Mundi VII*, 1950).

LEHMANN (Dr. E.). *Alte deutsche Landkarten*, 1935.

LOGA (V. v.). "Die Städteansichten in Hartman Schedels Weltchronick" (*Jahrbuch d. K. Preussischen Kunstsammlungen*, 1888).

MICHOW (A.). *Caspar Vopell und seine Rheinkarte vom Jahre, 1558*, 1903.

OBERHUMMER (E.). "Alte Globen in Wien" (*Anzeigphil.-hist. Akad. wiss.* Wien, 1922).

—— "History of Globes: a Review" (*Geogr. Review*, N.Y., 1924).

POPHAM (A. E.). *Georg Hoefnagel and the Civitates Orbis Terrarum*, 1936.

PRAESENT (H.). *Beiträge zur deutschen Kartographie*, 1921.

RAVENSTEIN (E. G.). *Martin Behaim: his Life and his Globe*, 1908.

RUGE (Dr. W.). "Aelteres kartographisches Material in deutschen Bibliotheken, 1904–06–11–1916" (*Nachrichten v. der Kgl. Gesellsch. d. Wissensch. zu Gottingen*).

RULAND (H. I.). "Survey of Double-Page Maps in 35 Editions of Seb. Münster" (*Imago Mundi XVI*, 1962).

SANDLER (C.). "Johann Baptista Homann" (*Zeits. d. Gesell. für Erdkunde zu Berlin*, 1886).

—— *Homann, Seutter und ihre Landkarten*, Amsterdam, Meridian Pub. Corp., 1964.

STOPP (Dr. K.). *Maps of Germany with Marginal Town Views*, Map Collectors' Circle, 1967.

WAGNER (H.). *Leitfaden durch den Entwicklungsgang der Seekarten XIII-XVIII Jahr.*, Bremen, 1895.

WEISZ (Prof. Dr. L.). *Die Schweiz auf alten Karten*, Zurich, 1945.

WOLKENHAUER (Dr. W.). *Erhard Etzlaubs Reisekarte durch Deutschland aus dem Jahre 1501*, 1919.

CHAPTER V

Holland and Belgium

THE Low Countries for roughly a century, from 1570 to 1670, produced in some respects the greatest map-makers of the world. The centres of production, at first in Antwerp and Duisburg, soon shifted to Amsterdam. For accuracy according to the knowledge of their time, magnificence of presentation and richness of decoration, the Dutch maps of this period have never been surpassed.

Early in the 16th century the cartographers of Belgium and Holland began to make their influence felt, and the movement began that finally wrested supremacy in map production from Italy and transferred it to Holland for a hundred years. Mercator was at work in 1534. Two years later, namely in 1536, Jacob van Deventer compiled two maps of Noord Holland and Brabant. These were not printed at that time, but in 1540 he executed a map of Holland, and this was first published in Mechlin in 1546 and again in Antwerp as a woodcut on nine leaves. He also compiled maps of Gelderland in 1542–3 and Friesland in 1545. These were copied and reproduced in Rome and Venice. Working during the same period, Cornelis Anthonisz drew two charts of the Zuyder Zee in 1541, and an important chart of the northern latitudes in 1543. This last became the standard map for navigation in the Baltic and Scandinavian waters for many years. It was eventually superseded by a more accurate map by Cornelius Doetz in 1589. The original has recently been found. There is a reprint by Visscher in 1610, and of this reprint only one copy is now extant, and is preserved in Oslo. It embraces Norway, the Baltic and White Sea area. Adrian and Hessel Gerritsz, Govert Willemsen and William Barentszoon were also responsible for remarkable sea charts.

It is impossible within a limited space, however, to give more than a slight mention of a few of the more interesting and valuable separate map publications. Ortelius, Mercator and Blaeu all published important single maps before the production of their atlases.

Abraham Ortel, or to give the better-known latinised form of his name, Ortelius, was born on April 4th, 1527, in Antwerp. He studied Greek, Latin and mathematics, and at the age of twenty set up as a map colourist and salesman, his sister aiding him. Business prospered, and he made friends with the literati of his own and other lands. In 1564 he published a map of the world on eight sheets (of which only one example is known, that in Basle University Library). This was followed in 1565 by a map of Egypt, and in 1567 by a large map of Asia on two sheets. In 1570 his greatest work appeared—the *Theatrum Orbis Terrarum*, or "Atlas of the Whole World." (78). The publication of this atlas marked an epoch in the history of cartography. It was the first uniformly sized, systematic collection of maps of the countries of the world based only on contemporary knowledge since the days of Ptolemy, and in that sense may be called the first modern atlas. (The 1513 Strassburg edition of Ptolemy was limited in scope, and contained classical as well as current maps, and the Italian collections that slightly preceded Ortelius were not uniform in selection, numbers, or format). It is said that the great Mercator delayed the publication of his own atlas in order that his younger friend Ortelius might have the honour of being first in the field with a standardised atlas. With the production of this atlas, initiative in map production may be said to pass from Italy to the Low Countries. The *Theatrum* was finely printed at the press of Christopher Plantin. Arnold Nicolai, Jacob van Deventer, Hendrik Terbuggen, Van der Borcht, Peter Draeck, Bernard van der Putte, Hans and Mynken Liefrink, Gilles Boileau de Bouillon, Hieronymous Cock, Christopher Sgrooten and Guicciardini all worked through Plantin.

The *Theatrum* achieved immediate popularity, four editions being printed in the first year (1570), and then edition followed edition, no less than 42 folio editions being printed between 1570 and 1612. The maps were also reduced and published in small format, and in this form reached 31 editions between 1576 and 1697. These varying editions were issued with the text in different languages—Latin, Dutch, German, French, Spanish, English and Italian; the maps being revised and added to or replaced by fresh surveys from time to time. The first issue was published on May 20th, 1570, and contained 70 maps on 53 leaves. To this nucleus were added five supplements or Additamenta in 1573 (17 maps), 1579 (23 maps), 1584 (23 maps), 1590 (22 maps) and 1595 (23 maps). As these Additamenta were published, they were incorporated with the atlas of that particular and subsequent years, so that the sum total of the maps was continually growing. Ortelius died on June 28th, 1598, and was buried at St. Michael's, in Antwerp.

Folio Editions of the "Theatrum Orbis Terrarum" of Ortelius

*1570 (Latin text)	*Additamentum II, 1579	
*1570 (Latin text)	Re-issued 1580	*1595 (Latin text)
*1570 (Latin text)	*1579 (Latin text)	*1598 (French text)
1570 (Latin text)	*1580 (German text)	*1598 (Dutch text)
*1571 (Latin text)	*1581 (French text)	1600 (Spanish text)
*1571 (Dutch text)	*Additamentum III, 1584	*1601 (Latin text)
*1572 (French text)	Latin and German, 1584;	1602 (German text)
*1572 (German text)	French 1585	*1602 (Spanish text)
1572 (German text)	*1584 (Latin text)	*1603 (Latin text)
*Additamentum I, 1573	1584 (Latin text)	*1606 (English text)
*1573 (Latin text)	*1587 (French text)	1607 (Latin text)
1573 (German text)	*1588 (Spanish text)	*1608 (Italian text)
*1574 (Latin text)	1589 (Latin text)	1609 (Latin text)
1574 (French text)	*Additamentum IV, 1590	*1612 (Latin text)
*1575 (Latin text)	1590 (Latin text)	*1612 (Italian text)
1578 (Latin text)	*1592 (Latin text)	*1612 (Spanish text)
1578 (French text)	*Additamentum V, 1595	1624 (Latin text)

The British Museum possesses all the editions in the above list marked with an (*).

De Jode

Gerard de Jode was born in 1509 in Nijmegen and died in Antwerp in 1591. He was eighteen years older than Ortelius, and started his cartographical career at an earlier date. From 1555 onwards de Jode published large engraved maps of Brabant, the World, Europe and Portugal, and a series of maps of Germany in 1569. De Jode evidently had in mind the idea of publishing a general atlas, but, less adroit or less influential than his younger rival Ortelius, he failed to secure a licence and monopoly for his work as the latter had done. At last, however, the atlas of de Jode appeared. It was entitled *Speculum Orbis Terrarum*, and was published in 1578. It consisted of two parts, the first containing 27 maps of various countries and districts, the second part, 38 maps of the provinces of the German Empire. The maps were

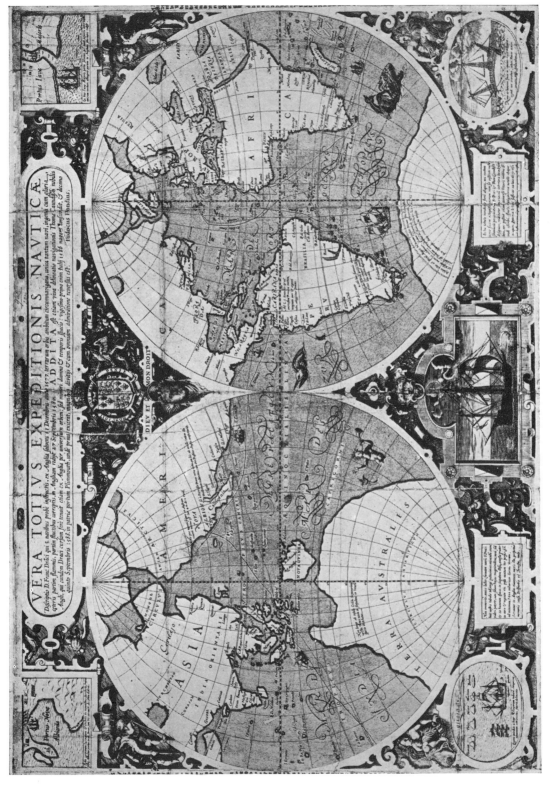

22   SIR FRANCIS DRAKE'S MAP OF THE WORLD (21¼" × 15")
Hondius, c. 1590. (*B.M. Maps MT 6 a2*)

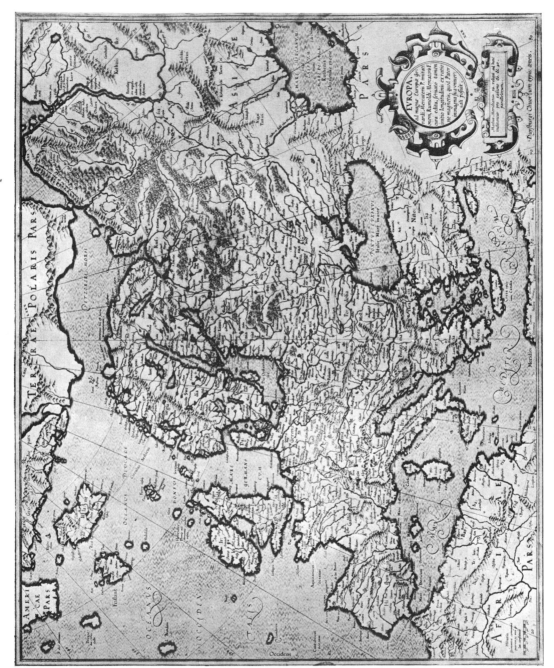

23 G. MERCATOR'S EUROPE (18½" × 15")
from Mercator's *Atlas*, 1595

finely engraved by the brothers Jan and Lucas van Doeticum. This first edition of the *Speculum* of Gerard de Jode is rare, not more than twelve copies are known. The rivalry between Ortelius and de Jode was evidently bitter, for Ortelius ignored de Jode in his list of authorities (though he used some of his material) and de Jode on his part abstained from mentioning Ortelius in his own list of cartographers. The maps of de Jode, though in some instances superior to those of Ortelius, were unable to compete effectively with those of his established rival. De Jode's atlas was revised and re-issued by his son Cornelius in 1593, when the total number of maps was increased to 83, this time the atlas appearing under the title *Speculum Orbis Terrae*. The merit of de Jode's work was not overlooked entirely by his compatriots, Hondius in particular bearing witness to his excellence.

Mercator (Gerard)

Gerard Mercator, the greatest name in geographical science after Ptolemy, was born in Rupelmonde on March 5th, 1512. At an early age he studied under Gemma Frisius, and his first work in the sphere in which he later became celebrated was the assistance he gave to Gemma Frisius in the construction of a Terrestrial and Celestial Globe in 1534–6. In 1537 Mercator started on his own account. His more important separate works produced prior to the issue of his atlas were:

1537	Map of Palestine	6 sheets	(1 example known, Biblioteca Civica di Perugia)
1538	Map of the World	—	(1 example known, American Geographical Society)
1540	Map of Flanders	4 sheets	(1 example known, Plantin Museum, Antwerp)
1541	Terrestrial Globe	—	(12 examples known)
1551	Celestial Globe	—	(9 examples known)
1554	Map of Europe	15 sheets	(1 example known, Stadtbibliothek, Breslau)
1564	Map of Gt. Britain	8 sheets	(3 examples known, Breslau, Rome, Perugia)
1569	Map of World	18 sheets	(Examples in Paris, Bib. Nat.; Breslau, Basle University and 1 private)
1572	Amended map of Europe	15 sheets	(3 examples known: Basle, Weimar, Perugia)

Van Ortroy cites a large map of America by Mercator on fifteen sheets. No copy is now known, and its existence has been doubted. Mercator's large maps were copied on a smaller scale by a host of imitators, not always with due acknowledgment. For example, Mercator's first world map of 1538 was copied by Salamanca, Lafreri and others; his map of Flanders by Tramezini, Stopius, Hogenberg, Guicciardini and Ortelius. His great world map was used in his own atlas on a reduced scale, both in general and in particular regions, and was copied or used as a base by Ortelius, de Jode and other geographers. In 1578 an edition of Ptolemy's *Geography* was published in Cologne, with maps constructed by Mercator. This Mercator-Ptolemy became extremely popular owing to the beauty of its engraving and the prestige of Mercator's name, and was re-issued in 1584, 1605, 1618, 1619, 1624, 1698 and 1704.

In 1585 appeared the best known of all Mercator's works: the first part of his celebrated *Atlas*, the word "atlas" being first chosen by Mercator to designate a collection of maps and only later adopted by all geographers (23). This *Atlas* contained three divisions, each with a separate title-page, France, Belgium and Germany, with 51 maps. The second part, Italy and Greece, appeared in 1590 with 23 maps; and the third and last part, posthumously in 1595 with 36 maps. Mercator died at Duisburg December 2nd, 1594.

Editions of Mercator's Atlas

*Part I of the *Atlas* 1585
*Part II of the *Atlas* 1590
*Part III of the *Atlas* 1595

*1595 The three parts issued together 107 maps
*1602 (Latin text, 274 leaves, including 111 maps)

Editions of Mercator's Atlas (continued)

*1606	Jodocus Hondius bought Mercator's plates and, adding 36 of his own, re-issued the atlas of Mercator (Latin text, 309 leaves, including 143 maps)		*1631	Appendix (99 maps)
			1631	(Latin text)
			1632	(Latin text)
			1633	(1st Dutch text)
			1633	(Latin text)
1607	(1st French text)		*1633	(French text)
1607	(Latin text) (146 maps)		*1633	Appendix (2 vols., 238 maps)
1608	(Latin text) (146 maps)		*1633	(1st German text)
1609	(2nd French text) (147 maps)		*1634	(1st Flemish text)
1611	(Latin text) (150 maps)		1635	(French text, Jansson Jodocus Hondius as publisher)
1612	(Latin text)			
*1613	(Latin text)		*1635	(1st English text) (191 maps)
1613	(French text)		*1635–7	(English text)
1616	(Latin text)		*1636	(English text) (195 maps)
1619	(French text) (156 maps)		1636	(German text, pub. Jansson)
*1619	(Latin text)		1637	Appendix (Flemish)
*1623	(Latin text) (156 maps)		1637	(Latin text)
1626	(French text)		1638	(Latin text)
1627	(Latin text)		1638	(*Atlantis nova Pars tertia*) (133 maps)
*1628	(French text)		1638	(Flemish text)
1628	(Latin text)		1636–8	(English text)
1629	(Latin text)		1638	Appendix (German)
1630	(French text)		1639	(English text)
*1630	Atlantis Appendix (60 maps)		1640	(German text)
*1630	(Latin text) (180 maps)		1641–2	(German text)

The British Museum possesses all the editions in the above list marked with an (*).

The atlases of Mercator were issued plain and hand-coloured, and their decoration is similar to that of Ortelius, but with a finer finish. His cartouches are generally somewhat smaller and more compact. A notable feature is the beautiful calligraphy employed on the surface of the maps.

Arnold Mercator, son of Gerard, born in 1537, compiled a map of the archbishopric of Treves and a plan of Cologne, also plans of his native Duisburg. In 1586 he commenced a map of Hesse, but he died in 1587 before its completion. Arnold Mercator had three sons—Gerard, John and Michael. Gerard the younger produced maps of Africa and Asia for the 1595 edition of the *Atlas*, based on his grandfather's materials, and in collaboration with his brother John was responsible for the engraving of many of the maps. John became geographer to the Duke of Cleves, and finished the map of Hesse begun by his father Arnold. Michael, the third son, composed the map of America for the 1602 edition of the atlas. Rumold, the youngest son of the first Mercator, saw the first completed edition of the *Atlas* through the press, to which he contributed the world map. He died in 1600.

Hondius (*J. and H.*)

The publishing business of Mercator passed eventually into the hands of Jodocus Hondius. Jodocus, born 1563, was first an engraver and secondly a publisher. At the age of twenty-eight he migrated to London and set up as a type founder and engraver. During his sojourn in England he married the sister of a compatriot who was also a map-maker—Pieter van den Keere (or Kaerius), who had engraved a set of maps of the counties of England in miniature, and Hondius, possibly through his influence, was employed later to engrave several of the maps in the folio edition of Speed's county atlas. Jodocus returned to Amsterdam before 1600 and, as stated above, acquired control of Mercator's stock (22). From 1606 he issued enlarged editions of Mercator's atlas. In 1608 Hondius produced a large scale world map and another

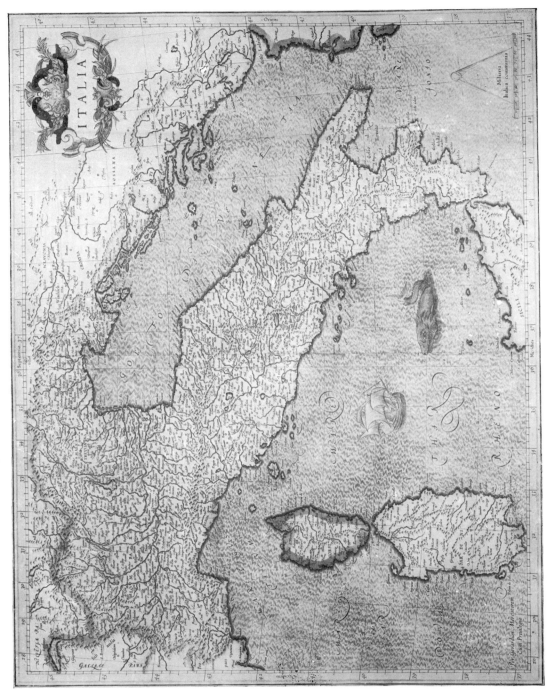

24 G. MERCATOR'S ITALY ($18\frac{1}{2}'' \times 14\frac{1}{2}''$)

from Mercator's *Atlas*, 1595

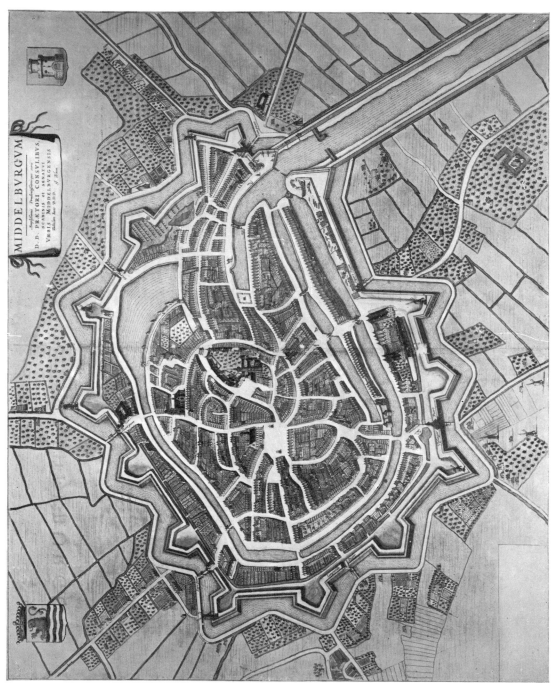

25 J. BLAEU'S PLAN OF MIDDELBURG (22½″ × 18¼″)
from Blaeu's *Atlas Maior*, 1662

in 1611. Jodocus died in 1611, and his son Henry succeeded to the business, and continued to issue still further improvements to the Mercator-Hondius atlas. In 1633 the name of Jan Jansson was joined with Henry Hondius in the latter's publication of the *Atlas*. Jan Jansson married the sister of Henry Hondius, and on the death of Henry about 1650 took over the business. Thus the impetus started by Mercator lasted for a century, and his projection, or the projection adopted by him, and the very term "atlas" itself passed into universal use, and are still with us.

To go back a little in time, towards the end of the 16th century B. Langenes produced a small atlas entitled *Caert Thresoor*. This was produced in Middelburg in 1598 with 165 maps by Hondius, van den Keer, Wright and Pigafetta, etc., re-issued in Amsterdam by Cornelius Claesz in 1599 and 1602 by De la Haye with the title *Thresor des Chartes*, and again in 1610. *Pieter Heyns* published the *Spieghel der Werelt* (P. Galle, 1577), and numerous later editions. The maps were reduced from Ortelius.

P. Plancius

The most important personality after Mercator at the close of the century was, however, Petrus Plancius. He produced an important planisphere in 1592, the second Dutch attempt at a large world map, and another in 1594. Dr. Wieder (*Monumenta Cartographica II*) gives a comprehensive list of no less than 80 maps compiled by Plancius. The work of Plancius is not so well known as that of many of his contemporaries, due to the fact that he did not issue an atlas. On the other hand, his material was extensively used by his contemporaries (19).

W. J. Blaeu

Willem Janszoon Blaeu was born at Alkmaar in 1571 and died at Amsterdam, October 21st, 1638. Willem Blaeu had two sons—Joan born in Amsterdam 1596, died in the same city 1673, and Cornelis, who died in 1642. The achievements of the firm of Blaeu were remarkable, and the palm for supremacy in map production in any age must be awarded to the Blaeus. Their work covered the whole range of Cosmography, Uranography, Hydrography, Chorography and Topography, the four branches having their beginnings as follows:

1599	Terrestrial and Celestial Globes.	1630	World Atlas *Atlantis Appendix*.
1608	Sea Atlas *Licht der Zeevaert*.	1644	General Atlas of Towns (25).

All these went into constantly improved and expanding editions.

The elder Blaeu, Willem, was a surveyor, globe-maker and publisher. Up to 1617 Willem Blaeu used his patronymic Janszoon, signing his work Guilielmus Janssonius or Willems Jans Zoon. This has sometimes led to confusion with his contemporary and rival Joannes Janssonius. His later work was signed Guilielmus or G. Blaeu. In 1604 Blaeu produced a map of Holland, and in 1605 a map of Spain, and a large world map. Of this world map only one example is known. It belongs to the New York Hispanic Society. In 1606 Blaeu produced another map of the world (an example being preserved in the Royal Geographical Society Library). Later a corrected edition was issued showing the Strait of Le Maire (discovered in 1616) and a further edition with the imprint "Janssonio" changed to "Blaeu." In 1608 he issued large maps of the four known continents—Europe, Asia, Africa and America. All these are extremely rare. Before 1630 he had issued 17 maps, and in 1629 he acquired the plates of 37 maps from Jodocus Hondius. From this small beginning sprang the great series of atlases that culminated in the *Atlas Maior*. The editions of the Terrestrial or World Atlases of Blaeu are as follows:

1629	Preliminary essay, 17 maps by Blaeu and 37 of Hondius.	
*1630	*Atlantis Appendix*, Amsterdam. (60 maps, no text).	
*1631	Appendix *Theatri A. Ortelii et Atlantis G. Mercatoris*, Amsterdam (117 maps and text).	
1634	*Novus Atlas* (161 maps, German text).	

1635 *Theatrum Orbis Terrarum sive Atlas Novus*, 2 vols. Four editions with text in Latin, Dutch, German and *French.
1640 *Theatrum*, 3 vols. (Four editions, *Latin, Dutch, *German, *French).
1640–5 *Theatrum*, 4 vols. (Four editions *Latin, Dutch, *German, *French).
1648–54 *Theatrum*, 5 vols. (*Latin, German and French).
1648–58 *Theatrum*, 6 vols. (Latin, German, French, Spanish, *Dutch).
1662 *Atlas Maior* (Latin text, 11 vols.).   (French text, 12 vols.).
          (Dutch text, 9 vols.).   *(Spanish text, 10 vols.).
          (German text, 9 vols.).

Those editions marked with an (*) are to be found in the British Museum.

In 1633–4 Blaeu was appointed map-maker to the Dutch East India Company. In 1648 J. Blaeu compiled a large map of the world—*Nova totius Terrarum Orbis Tabula* on twenty sheets. One copy is preserved in the Royal Geographical Society Library and another in the British Museum. For this magnificent production Blaeu searched the records of all nations, the map bearing names and annotations in Dutch, English, Latin, French, Spanish, etc., and it was revised practically up to the day of publication. Wieder justly calls it "the highest expression of Dutch cartographical art."

J. Jansson

The contemporary and rival of Blaeu was Jan Jansson, born 1596 in Arnheim, and died in Amsterdam in 1664. Jansson produced maps of France and Italy in 1616, and issued an edition of Ptolemy's *Geography* in 1617. He likewise constructed globes. In 1633, in conjunction with his brother-in-law Hondius, he issued the second volume of the *Mercator-Hondius Atlas*.

His works include:

1638 *Nieuwen Atlas ofte werelt Beschrijvinge*. 2 vols.
1639 *Theatre du Monde*. 3 vols. French text.
1640 3 vols. German text.
1642 French text in conjunction with Hondius, 3 vols.
1642 3 vols. Dutch text. Reissued 1645.
*1647 4 vols. Latin text [vol. 4 being Great Britain].
*1649 4 vols. German text.
1653 4 vols. Spanish text. Reissued 1666.

1650 The 5th volume appeared, the *Sea Atlas*, also the Ancient World. Reissued 1652.
1656 6 vols. Latin text. *Reissued 1658.
*1658 6 vols. German text.
*1647–58 6 vols. French text.
1647–62 10 vols. Dutch text.
1658–61 11 vols. Latin text.
*1666 *Atlas Contractus*. 2 vols.
1657 *Illustriorum . . . Urbium*. 8 vols.

The British Museum possesses examples of those marked with an (*).

Jansson's business was acquired by Peter Schenk, who republished his atlas about 1683. Schenk also issued an atlas *Hecatomopoli*, of 100 plates, in 1702, and an *Atlas Contractus* about 1709. Later he was associated with Leonard Valck, and Schenk and Valck between them issued an atlas of English county maps.

Another important family of map publishers were the Visschers, who worked all through the 17th century and into the first years of the 18th (27). The eldest member of the family was Claes or Nikolaus J. Visscher (1587–1652), who was succeeded by Nikolaus (1618–79) and Nikolaus (1649–1709). They produced atlases on their own account, an *Atlas Contractus* about 1660 and an *Atlas Minor* about 1680, which went into several editions, and their plates were used extensively by other map-makers both in Holland and abroad. The work of the elder Visscher is uncommon and very pleasing, similar in character to that of Blaeu, though not of quite so high a quality.

Other Dutch map publishers at the turn of the century were Allard (Atlas Major, 3 vols., fol., *c.* 1710), Danckerts and Ottens (the last named issuing a large *Atlas Maior* in seven folio volumes with 835 maps), all of whom issued atlases (73). These were followed in the 18th century by Covens and Mortier, Pieter van der Aa and Isaac Tirion. Covens and Mortier issued several atlases, largely based on the work of Sanson and De l'Isle, their most ambitious effort being their *Atlas Nouveau* in 9 folio volumes, published over a period of years from 1711 to 1760. They also issued a series of large-scale maps of the world and the four known continents about 1705 on twenty-five sheets. Pieter van der Aa issued an even more elaborate work, his *Galerie Agréable du Monde* running to 66 volumes, with about 3,000 plates. This was issued in 1729 and limited to 100 sets. He also issued an *Atlas Nouveau et Curieux* in 1714, with 139 maps, and an *Atlas Nouvel* in the same year, with 98 maps. But in spite of this quantitative output, Dutch cartography had by this date suffered a decline. The works of the later Dutch publishers were compilations and imitations. Initiative had passed to France, and French theories and French science dominated the 18th century.

To return to the 17th century, contemporaneously with the large output of terrestrial or land atlases, a highly important, valuable and lesser-known series of Navigation or Sea Atlases was turned out from the Dutch presses. Their scarcity is due partly to the fact that fewer editions and probably smaller quantities were printed (their interest would practically be confined to maritime powers) and apparently they did not arouse an extensive academic interest as was the case with the land atlases. The following is a brief, but not comprehensive, list of Dutch Marine works:

Alphen (Pieter van). *Nieuwe zee atlas ofte water weerelt*. Rotterdam, fol., 1660 (12 charts).
—— —— Other editions 1660 (15 charts), 1682 (14 maps).
Barentszoon (Wm.). *Caert boeck van de Midlandtsch Zee.* Amsterdam, 1595 (9 charts). The first Dutch maritime atlas of the Mediterranean. Reissued 1600, 1605, 1606, 1609, 1626.
—— —— French translation, *Description de la Mer Mediteranée.* Amsterdam, 1608. Reissued 1626.
Blaeu (Willem Jansz). *Licht der Zeevaert.* Willem Jansz, 1608, 1610, 1613; other editions 1617, 1618, 1620, 1621, 1623, 1627, 1629, 1630, 1634, 1646.
—— French translation, *Flambeau de la Navigation*, 1619; other editions 1620, 1625.
—— French translation, *Phalot de la Mer*, J. Jansz, 1619, 1620, 1625. Reissued 1637.
—— English translation, *Light of Navigation*, 1612; other editions 1620, 1622 and 1625.
—— *Zeespieghel*, fol., 1623; other editions 1624, 1627, 1631, 1638, 1640, 1643, 1650, 1652, 1655 and 1658.
—— English translation, *Sea Mirror*, 1625 and 1635; *Sea Beacon*, 1643 and 1653.
Carolus (Jovis). *Het nieu vermeerde Licht.* Amsterdam, Jansson, 1634.
Colom (Arnold). *Zee Atlas.* Amsterdam, 1656 (15 maps).
—— *Atlas of werelts water deel.* Amsterdam, 1660–61 (23 maps).
—— *Zee Atlas maritimo o mundo aquatico.* Amsterdam, 1669 (52 maps).
Colom (J. A.). *De Vyerighe Colom*, 1632–3. French edition by Bardeloos, 1633. English edition (*Upright Fyrie Colomne*), 1648.
—— *De Groote Lichtende ofte Vyerighe Colom*, 1652, 1661, 1663.
Doncker (H.). *De Zee-Atlas of Water-waereld*, 1659 (18 charts); other editions 1660–1 (27 charts), 1663 (26 charts), 1665 (30 charts), 1666 (32 charts).
—— *Atlas del Mondo o el Mundo Aguado.* Amsterdam, 1665 (46 charts).
—— *De Nieuwe groote vermeerderde Zee Atlas*, 1674 and 1676 (27 charts).
—— *Nieuw groot Zee Kaert-boeck*, 1712 (28 charts).
—— *Nieuw en groote Lootsmans Zee Spieghel*, 1661–6 (30 charts).
Gerritsz (Adr.). *De Zeevaert O. ende W. Zee vaertwater*, Amsterdam, C. Claes, 1588.
—— (Hessel). *Besch. vander Samoyeden Landt.*, 1612. Latin editions, 1612 and 1613.
Goos (A.). *Nieuw Nederlandtsch Caertboeck.* Amsterdam, 1616 (23 charts) and 1625.
Goos (P.). *De Lichtende Colomme ofte Zee spiegel.* Amsterdam, 1650, 1654; other editions, Amsterdam, 1657 and 1664.
—— *De Nieuwe Groote Zee-spiegel.* Amsterdam, 1662 (84 charts); also Amsterdam, 1675 (65 maps).

—— *De Zee Atlas ofte Water weereld.* Amsterdam, 1666 (40 charts); other editions, Amsterdam, 1666 (41 charts), 1668 (41 charts), 1669 (41 charts), 1670 (41 charts), 1672 (41 charts), 1676 (40 charts). English translation, *Sea Atlas or the Watter-World*, 1668 (41 charts), 1670 (40 charts).

—— *Le Grand et nouveau Miroir ou flambeau de la Mer.* Amsterdam, 1671 (33 charts).

Haeyen (A.). *Amstelredamsche Zee-Caerten.* Leyden, Plantin, 1585 (5 charts); other editions 1591, 1605, Amsterdam, Corn. Claesz, 1609 (5 charts), Amsterdam, Dirck Pietersz, 1613.

Hooge (Romain de). *Atlas Maritime.* Amsterdam, 1693 (38 charts).

—— *Zee Atlas tot het gebruik van de vlooten des Konigs van Groot Britanje,* 1694.

Jacobsz (T.). *Nieuw groot Straetsboeck inhoudende d' Middellantse zee,* 1648 (20 charts).

—— *'t Nieuwe en vergroote zee-boeck.* Amsterdam, 1653.

—— *Lightning Columne or sea mirror.* Amsterdam, C. Lootsman, 1668 (60 charts); and 1676 and 1689.

—— *Nieuw en Groot Zee Spieghel.* Amsterdam, J. and C. Lootsman, 1669 (24 charts); Amsterdam C. Lootsman, 1689.

—— *Lighting Colom of the Midland Sea.* Amsterdam, C. Lootsman, 1692.

—— *Nieuwe groote geostroyeerde verbeterde en vermeerderde Lootsman zee Spigel,* 1707.

Jansson (J.). *De Lichtende Columne ofte Zee spiegel,* 1651, 1652.

Keulen (J. van). *De Groote nieuwe vermeerderde Zee Atlas ofte water werelt.* Amsterdam, 1695.

—— *Zee Atlas,* 5 vols., from 1682–1684 (Bom). 1 vol., 1681, 31 charts (Tiele). 1 vol., 1688, 34 charts (Haan). 1 vol., 1694, 146 charts (Muller).

—— *De Lichtende Zeefakkel,* 1681–96 (L.C.) (78 charts).

—— *La Grande Ilumminante Nueva Antorcha del Mar.* Amsterdam, 1698.

—— *La Nueva y Grande Relumbrante Antorcha del Mar.*

—— *Do groote nieuwe vermeerded zee atlas,* 1736.

Loon (J. van). *Klaer lichtende noort ster ofte zee atlas.* Amsterdam, 1661 (46 charts), 1666 (47 charts), 1668 (49 charts).

Mortier (P.). *De fransche neptunus,* 1693–1700.

Ottens (R. & J.). *Atlas van Zee vaert.* Amsterdam, 1745 (33 charts).

Robijn (J.). *Zee Atlas,* 1683 (20 charts), 1683 (40 charts).

—— English edition, *The New Enlarged Lighting Sea Column,* 1689 (137 charts).

Roggeveen (A.). *Het eerste deel van het Brandende Veen, verlichtende alle de vaste kusten ende eylanden van geheel West-Indien ofte rio Amazones.* Amsterdam, P. Goos, 1675 (33 charts). (And 1676.)

—— —— Spanish edition: *La primera parte del monte de turba ardiente allumbrado con la claridad de su fuego todos los costas firmes, y ysles de toda la India Occidental.* P. Goos, 1680 (33 charts). Also French and English texts in 1676.

Vooght (C. J.). *Nieuwe groote lichtende zee fakkel,* 1782.

Waghenaer (Lucas Jansz). *Teerste Deel van de Spieghel der Zeevaerdt,* 1584 (23 charts).

—— *Het tweede Deel van de Spieghel der Zeevaerdt,* 1585 (added to Vol. I, making 44 charts).

—— *Spieghel der Zeevaerdt.* Both parts dated 1585.

—— *Speculum Nauticum* (Latin text), 1586 (45 charts). (Dutch text), 1588 (46 charts). (Latin text), 1591 (47 charts). (French text), 1590. (Dutch text), 1596.

—— *Mariners' Mirrour.* A. Ashley, 1588 (45 charts).

—— *Den Nieuen Spieghel der Zeevaerdt van de Navijatie der westersche Zee,* 1596. French edition, 1605.

—— *Thresour der Zeevaert,* 1592, 1596, 1598, 1602 and 1608. French edition, 1601.

Willemsen (G.). *Die caerte vande Oost ende West Zee,* 1588.

Wit (F. de). *Zee Karten.* Amsterdam, 1675 (27 charts).

AUTHORITIES

AMSTERDAM. *Collections-Mensing au Scheepvaart-Museum* (Musée Maritime), 2 parts, Amsterdam, F. Müller, 1923–6.

ANDREAE (S. J. Fockema) and B. VAN'T HOFF. *Geschiedenis der Kartografie van Nederland,* 1947.

AVERDUNK (H.) and Dr. J. Müller REINHARD. *Gerhard Mercator und die Geographen unter seinen Nachkommen,* 1914.

AVIS (J. G.). *Jacob van Deventer Kaart van Gelderland van* 1556. Gels. Bijdr. an Mededselingen, 1935.

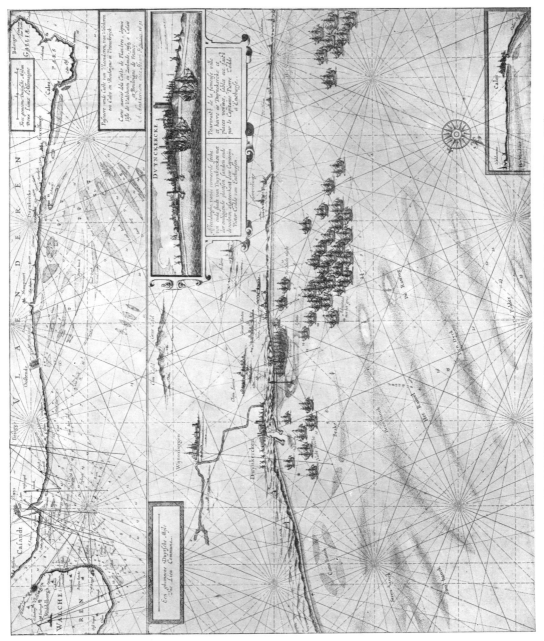

26 H. HONDIUS'S CHART OF DUNKIRK ROADS, 1631 (21" × 17")

27 C. I. VISSCHERS OVER-YSSEL, 1660 (20″ × 15½″)

BAART DE LA FAILLE (R. D.). "Nieuwe Gegevens over Lucas Janszoon Wagener'" (*Mariners' Mirror*, 1937).

BAUDET (P. J. H.). *Leven en Werken van Willem Jansz Blaeu*, Utrecht, 1871.

BRUUN (C. G.). "Cornelius Antoniades Kaart over Denmark og Zeinkaartet of 1558" (*Geogr. Tidskrift*, Copenhagen, Vol. IX, 1887–8).

BURGER (C. P.). "Oude Hollandsche Zeevart-Uitgaven" (*Het Boek*, 1913).

—— "Oude Zeekaartboeken in nieuwe uitgaven" (*Het Boek*, 1917).

D'AILLY (A. E.). *Catalogus van Amsterdamsche Plattegronden*, Amsterdam, 1934.

DENUCE (Jan). *Oud Nederlandsche Kaartmakers in betrekking met Plantin*, 1912–13.

DE SMET (A.). "A note on the Cartographic work of Pierre Pourbus, Painter of Bruges" (*Imago Mundi IV*, 1947).

—— "Une carte très rare Le Gallia Belgica de Gilles Boileau de Bouillon" (*Rev. Belge de Phil. et d'Hist.*, Bruxelles, 1939).

DERMUL (A.). "Les globes terrestre et céleste de W. Jansz Blaeu conservés à la Bib. Comm. d'Anvers" (*Gaz. Astron.*, Anvers, 1940).

DESTOMBES (M.). *Cat. des cartes nautiques MS. sur parchemin 1300–1700. Cartes Hollandaises. La Cartographie de la Compagnie des Indes Orientales 1593–1743*. Saigon, 1941.

—— *La Mappemonde de Petrus Plancius gravée par Josua van der Ende 1604*. Tonkin, 1944.

GERARD (P. M. N.). "Les globes de Guillaume Blaeu" (*Soc. Geogr. d'Anvers Bull.*, 1883).

HEAWOOD (E.). "A Masterpiece of John Blaeu" (*Geogr. Jnl.*, Vol. LV, 1920).

—— "An unrecorded Blaeu World Map of c. 1618" (*Geogr. Jnl.*, Vol. CII, 1943).

HESSELS (S.). *Abraham Ortelii … epistulae*, London, 1887.

HOFF (B. van't). *De Karten van de Nederlandsche Provincien in de zestiende Eeuw door Jacob van Deventer*, s'Gravenhage, 1941.

HONDIUS (I.). *Map of the World on Mercator's Projection 1608*, with memoir by E. Heawood.

JACOBSZ (G.). "Uber die Meile bei Lucas Jans Waghenaer 1583 und zugehörige Fragen" (*Tijdsch. K. Nederl. Aard. Gen.*, Amsterdam, 1938).

KEUNING (J.). "Jodocus Hondius Jr." (*Imago Mundi V*, 1948).

—— "Hessel Gerritsz" (*Imago Mundi VI*, 1949).

—— "Cornelis Anthonisz" (*Imago Mundi VII*, 1950).

KOEMAN (Cornelis). *Atlantes Neerlandici*. 5 vols, 4to, Amsterdam, 1967–71.

LYNAM (E. W. O'F.). "Floris Balthasar, Dutch Map Maker, and his sons" (*Geogr. Jnl.*, 1926).

—— "Lucas Waghenaer's Thresoor der Zeevaert" (*B.M. Quarterly*, Vol. XIII, 1939).

MARINE (Ministerie van). *Catalogus der Verzameling van Kaarten van het ministerie van Marine*, 's Gravenhage, 1872.

ORTROY (F. van). *Bibliographie Sommaire de l'Oeuvre Mercatorienne*, 1920.

—— *G. de Jode et son Œuvre*, 1914.

—— *Remarkable Maps of the 15th, 16th and 17th Centuries*, Amsterdam, Müller, 1894–7.

RIJKS-ARCHIEF. *Inventaris der Verzameling Kaarten berustende in het Rijks-Archief*, 's Gravenhage, 1867.

RUGE (Dr. W.). *Aelteres Kartographisches Material in deutschen Bibliotheken*, 1904–16.

STEVENSON (E. L.). *Willem Janszoon Blaeu, 1571–1638, a sketch of his life and work with especial reference to his large world map of 1605*, New York, 1914.

—— and J. FISCHER. *Map of the World by Jodocus Hondius, 1611*. New York, 1907.

TIELE (P. A.). *Mémoire bibliographique sur les journaux des navigateurs Néerlandais*, Amsterdam, 1867.

—— *Nederlandsche Bibliographie*, Amsterdam, 1884.

TOOLEY (R. V.). *Leo Belgicus, a List of Variants*. Map Collectors' Circle, 1964.

VAN RAEMDONCK (J.). *Gerhard Mercator, sa vie, ses œuvres*, 1869.

WAUWERMANS (Lt.-Gen.). *Hist. de l'Ecole Cart. Belge et Anversoise du XVI siècle*, Bruxelles, 1895.

WIEDER (Dr. F. C.). *Monumenta Cartographica*, 1925–34.

—— *Dutch Discovery and Mapping of Spitsbergen 1596–1829*, Amsterdam, 1919.

—— *Nederlandsche Hist.-Geogr. Documenten in Spanje*, Leiden, 1915.

CHAPTER VI

French Cartography

FRANCE had schools of geography at an early date in Paris and elsewhere, but the disturbed state of the country was not conducive to prolonged effort, and French contribution to cartography did not become really effective till a comparatively late date. Sacrobosco expounded cosmography in Paris in the 13th century, Cardinal d'Ailly speculated in the 15th century, and Waldseemüller worked at St. Dié in Lorraine on the 1513 edition of Ptolemy. This was printed in Strassburg, and the first edition of Ptolemy to be printed in France was that by Gaspar and Melchior Treschel in Lyons, 1535. Gaspar produced a second edition at Vienne in Dauphiny in 1541.

Mention may be made of the school of portolan makers at Arques, based on the labours of the mariners of Havre and Dieppe, of whom the most celebrated were Desliens (1541), Desceliers (1546), Vallard (1547) and Le Testu. Of the last-named a MS. atlas is in existence containing no less than 7 world maps, all on different projections.

But the most prominent French cartographer in the 16th century was Oronce Finé (Orontius Finæus). Born in Briançon in 1494, he studied in Paris and died in 1555. He produced a heart-shaped map of the world in 1519, which he dedicated to Francis I. This was copied by Apian in 1530. In 1531 he drew a map of the world for the work of Grynæus, *De Novus Orbis*: this was on a double heart-shaped projection. It was published in Paris in 1532, and in another edition by Münster in Basle. This map was copied and made use of by many geographers, including the great Mercator himself. In 1534 Finé re-drew a single heart-shaped projection of the world, and this was engraved in 1536. Two examples are known, one in Paris and one in Nürnberg. This map secured a wide circulation through the copies made in Italy by Lafreri in 1556 and 1566. A final world map by Finé appeared in 1544, but no copy is now known. He likewise constructed a map of France in 1525 (a woodcut on four leaves), and this map was extensively copied on a reduced scale in Italy. In 1544 he drew a map of Dauphiny, a map of France, and as already stated a world map. None has survived.

Nicolas de Nicolay, sieur d'Arfeville et de Bel Air, cosmographe du Roi, was born in 1517 in Dauphiny. He constructed a navigational map of the coasts of Europe and North Africa in 1544, which was printed on four copper-plates. Nicolay was responsible for the marine map of the New World in Medina's *Art of Navigation*, published in Lyons in 1554, and this map was reproduced by Camocio in Venice in 1560. He also drew a map of the Boulonnois, Guines and Calais area, a large-scale six-leaved map, of which the only known copy·is preserved in Paris. His other works include 7 maps and plans of the Duchy of Berry and the Baillage de Bourges, a map of the Duchy of Bourbonnois dedicated to Charles IX

ATLAS
NOUVEAV.
CONTENANT TOUTES LES
PARTIES DU MONDE,
OU
Sont exactement Remarqués
LES EMPIRES, MONARCHIES,
ROYAUMES, ESTATS,
Republiques & Peuples qui
s'y trouuent à present.
PAR LE S*. SANSON
Geographe ordinaire du Roy.
PRESENTÉ
A MONSEIGNEUR LE DAUPHIN
Par son tres-humble, tres-obeissant,
et tres-fidéle Seruiteur,
HUBERT IAILLOT
Geographe du Roy.

A AMSTERDAM
Chez PIERRE MORTIER, et Compagnie
Avec Privilege de Nos Seigneurs les Estats

LOUIS LE GRAND ROY DE
FRANCE ET DE NAVARRE.

L'HERCULE
FRANÇOIS.

MONSEIGNEUR LE DAUPHIN.

MARS

PALLAS

BATAILLE DE CASSEL.

BATAILLE DE SENEF.

28   TITLE-PAGE OF JAILLOT'S *ATLAS NOUVEAU*, c. 1700 (14″ × 18″)

29 PLANCIUS'S MAP OF FRANCE (19″ × 15¾″)
from the English edition of Ortelius's *Theatre of the Whole World* 1606

and Catherine de Medici in 1569, and a map of the coasts of Scotland, the Orkneys and Hebrides, which was printed in Paris in 1583.

Other geographers of 16th-century France were Jean Jolivet, geographer to François II, who composed a large map of Berry on six leaves for Marguerite of Navarre in 1545, and a two-sheet map of Normandy in the same year. He likewise completed a survey of Picardy (undated), and a map of the whole of France on four woodcut leaves in 1560, entitled "Nouvelle Description des Gaules" (only known example preserved in Breslau).

André Thevet, born 1502 in Angoulême, travelled extensively when young in Europe and America. He produced separate maps of the four continents, a map of France and a world map in the shape of a fleur-de-lis (the two latter are lost and known only from literary reference).

Gilles Boileau de Bouillon included a map in his description of Germany 1551 copied from Münster, and 4 small maps, including Peru, in a work entitled *Le sphere des deux mondes*, 1555. A map of Savoy on two copper-plates appeared in 1556, and this was re-issued in Italy and also used by Ortelius. His final works were *La Campagne de Rome* (known from a single copy at Wolfegg), which was extensively copied in Italy, and a map of Belgium, the original of which is lost; but his representation is preserved in later Italian copies under the title *Gallicum Belgicum descripsit*.

Charles de l'Escluse, better known under the latinised form of his name Carolus Clusius, was born in Arras in 1526. A naturalist, he was largely responsible for the propagation of the potato in the 16th century. He produced two maps, *Gallia Narbonensis* and a large map of Spain on six leaves (known from a single copy in Basle).

Gabriel Symeone, a Florentine resident in France, compiled a map of Auvergne in 1560. This was printed in Tours by Guillaume Rouille, and was used by Ortelius, Bouguereau and Blaeu.

Jean Chameau, Advocate of Bourges, compiled a map *Carte du Berry*, Lyons 1566, which was reproduced by Ortelius and Bouguereau.

Ferdinand de Lannoy, comte de le Roche, military engineer, born 1542, compiled in 1563 a map of Bourgoigne and in 1565 a map of Franche Comte.

Jean and Jacques Surhon, born in Mons, issued 4 maps, Namur in 1555 used by Ortelius, Bouguereau, Blaeu, Jansson and Visscher; Vermandois in 1558, printed by Plantin, but no copy known to survive; Picardy, which remained in manuscript until used by Ortelius in 1579 and later by Bouguereau, Kaerius and others; and finally Artois, which was likewise extensively used by contemporary publishers.

All the preceding maps, with the exception of the printed editions of Ptolemy, are extremely rare, known only from single or very few examples, and some from reference only. None is likely to come the way of collectors, except very occasionally the later Italian copies produced in Rome and Venice. But from 1570 onwards, that is, from the first atlas of Ortelius, the output of maps increased enormously, edition after edition poured from the presses in the Low Countries (see Chapter IV), and it is easy to procure single maps of the end of the century by Ortelius, Mercator and Hondius.

The first series of maps of the provinces of France appeared in the *Theatrum* of Ortelius in 1570, seven provinces only being issued in this edition. In 1579 Ortelius increased the number to ten, the additions being Poictou by Pierre Roger, Anjou by L. Guyet, Picardy and Artois by Surhon, and Burgundy by de Lannoy. Five further maps were added in 1590: Maine by M. Oger, Provence by J. P. Bompare, Lorraine by himself, and Burgundy, Brittany and Normandy without indication, making 17 maps in all. De Jode, in his atlas of 1578, likewise gave 7 maps of the provinces of France, and in the edition published by his son in 1593, maps of Artois, Maine and Venaissin were added. In 1585 Mercator issued his *Galliae Tabulae Geographicae*, including a general map and 11 maps of the provinces of France.

It was on the foundation provided by these maps that Maurice Bouguereau produced the first national atlas of local French cartography. It was issued at Tours in 1594 under the title of *Le Théâtre François*. This formed the basis of following national atlases of France, the nearest equivalent to Saxton's atlas in English cartography. It consists of general maps (varying from one to three in different copies) and 15 maps of French provinces. They were engraved by Tavernier. For 3 of the maps, Blaisois, Touraine and Limoisin, it is the first printed representation. The general maps are by Petrus Plancius (29), Jean Jolivet or Guillaume Postel. The atlas is rare: Fordham lists only six copies as being known to him, viz., in the British Museum, Bib. Nationale Paris, Tours, Dresden and two in his own possession (later donated to the British Museum). Since Fordham's time, three further copies have appeared, and others may yet come to light.

Jean Le Clerc took over the maps of Bouguereau and re-issued them under the title *Théâtre Géographique du Royaume de France*.

Fordham gives the following:

1617   A possible edition.
1619   A unique copy in the British Museum.
1620   A copy in Paris Bib. Nationale. 39 maps.
1621   Maps increased to 45 (copy in Univ. Lib., Cambridge).
1622   Issued by his widow. 49 maps.
1626   Ditto.
1631   Number of maps, 52.

The publication was finally taken over by J. Boisseau (enlumineur du roi pour les cartes géographiques), who issued it under a new title, *Théâtre des Gaules*, in 1642, with 75 maps.

In 1634 Melchior Tavernier likewise issued an atlas with the same title as that of Le Clerc, viz. *Théâtre Géographique du Royaume de France*. There are two examples in the British Museum, one containing 80, the other 95 maps. A second edition appeared in 1637. The maps of Bouguereau are practically unobtainable and the others are uncommon, but it is possible to pick up some examples of Tavernier maps from time to time (31).

The 17th century commences with a fine large-scale map of France by François de la Guillotière, wood-cut on nine sheets, presented to Louis XIII in 1612 or 1613, though commenced in 1596. This was issued by Le Clerc and re-issued in a new edition by his widow in 1632.

We now come to the great school of French geographers initiated by Nicolas Sanson, who in effect laid the foundations of what is known as the French school, the beginning of a great period, for from the time of Sanson in the second half of the 17th century till the latter part of the 18th French geographical conceptions were dominant, the centre of map production shifting from the Low Countries to France.

Nicolas Sanson, born in Abbeville in 1600, died in Paris in 1667. He had three sons, Nicolas (who died in 1648), Guillaume (died 1703) and Adrien (died 1708), and a grandson, Pierre Moulard Sanson. Sanson published about 300 maps. His first cartographical work is said to be a map of ancient Gaul on six sheets, printed in 1629, though composed at an earlier date when he was only eighteen years old. His first atlas was issued in 1654 with 100 maps. A second atlas appeared in 1658 with a printed title *Cartes générales de toutes les parties du monde* and a printed Contents leaf listing 113 titles. This was published by Pierre Mariette. Another edition appeared in 1664, also published by Mariette, with a printed list of 147 titles, and the Library of Congress lists a copy in their possession in two volumes of 1670 with 153 maps. The atlases of Sanson are not often found in their prescribed state; it would seem a fairly common practice for additional maps to be inserted, as the number rarely seems to agree with the printed list. Also many copies of Sanson's atlases now preserved contain maps by other cartographers, and many that can be ascribed only doubtfully to Sanson.

After Sanson's death in 1677 another edition was issued by his sons Guillaume and Adrien (in the British Museum). This copy has 102 maps. The Sansons also issued a series of four 8vo volumes on the four continents, as follows:

L'Europe en plusieurs cartes nouvelles. Paris, 1648 and 1651.
L'Afrique. Paris, 1656 and 1660.
L'Asie. Paris, 1652, 1653, 1658 and 1662.
L'Amérique. Paris, 1656, 1657, 1662 and (1676),

the first by the younger Sanson, the other three by the elder. Sanson's maps are not nearly so ornamental as the maps of the Dutch school. Their embellishment is confined to a title cartouche for each map, but they are very clear and neat, pleasing to the eye and of handy format (average size 20 by 16 inches). Sanson used Tavernier, Mariette, Peyrounin, Cordier and Jean Somer as publishers and engravers, among them Cordier being the finest artist.

The Sansons were succeeded by Jaillot as the most influential figure in map production in France. Alexis Hubert Jaillot was born in Franche Comté about 1632 and died in Paris 1712. Originally a sculptor, he married the daughter of Nicolas Berey (30), a renowned map colourist (enlumineur de la reine), and becoming enamoured of his father-in-law's profession, he took up the study of geography, collaborating with the younger Sansons and succeeding to the greater part of their maps and plans. Jaillot had them engraved on a larger scale with the greatest care and exactitude, and they form the basis of the greater part of the atlases issued by him, though Jaillot himself constructed some maps and plans of fortresses. Jaillot's maps are in large and handsome format, finely printed on the best paper, the titles and scales of miles within elaborate and large cartouches, usually depicting the characteristic costumes and products of the country delineated, and some have inset views. In ordinary examples the maps themselves were coloured in outline and the cartouches left plain, but copies for special clients were fully coloured and illumined with gold. Sir George Fordham calls such examples "the finest specimens extant of this decorative art," Jaillot published an *Atlas Nouveau* in 1681, which contained an attractive title frontispiece and 45 maps (28). Other editions appeared in 1689, 1691 and in 1692 with a dated title. A further edition appeared in 1695, with a fine portrait of Jaillot, an engraved title changed to *Atlas François* and 115 maps. The venture was popular, and continued to be issued for many years, that of 1696, edited by Pierre Mortier, had 136 maps and 196 plans on 28 sheets. Another copy, 1700–50, with 167 maps. Jaillot also issued *Le Neptune Français ou Atlas Nouveau des cartes marines* in 1693. Other members of the Jaillot family were Bernard Jean Hyacinthe, Bernard Antoine, and Chauvigne Jaillot, the family continuing in operation till 1780.

Other geographers at the end of the 16th century were:
Pierre Duval, the son-in-law of Sanson (1619–83), who published the following works:

Les acquisitions de la France par la paix. Paris, 1658. Second edition, 1667, with 14 maps and 3 tables.
Cartes de géographie, 12mo, 1662, with 40 maps.
Divers cartes et tables pour la Géogr. ancien, 8vo, 1665, with 44 maps. (Reissued 1669.)
Cartes et tables de géographie des prouinces eschués, 12mo, 1667, with 10 maps.
Le Monde ou la Géogr. Universelle, 12 mo, 1670, with 84 maps. (Reissued 1682.)
Les XVII prouinces où sont les conquestes du Roy en Holland et en Flandres. Paris, 12mo, 1679, with 18 maps.
Monde chrestien, 12mo (1680), with 16 maps.
(*Géographie Univ.*, 1685), 12mo, with 51 maps. Second edition London, 1691, and in Nürnberg, 1694.
Atlas of 102 maps published by his daughter 1688–9, folio.

And Nicolas de Fer, born 1646, died 1720, engraver and geographer. He had a prolific output, his ingenious ornamentation being more successful than his geographical exactitude. His productions are:

La France triomphante sous le règne de Louis le Grand, 6 sheets, 1693. This has over 200 cartouches.
Les Forces de l'Europe ou descriptions des principales villes, 1696, with 177 maps, 8 plates and a table. Issued in 8 parts, each with a title-page with different dates ranging from 1693 to 1697. Re-issued by Mortier in Amsterdam about 1702 and by Van der Aa with additions in 1726.
Plusieurs cartes de France avec les routes et le plan des principales villes, Paris, 1698, 1726, 1730, 1756, 1760 and 1763.
Atlas royal, 1699–1702, folio, with 86 maps.
Atlas curieux, 1700–1703, oblong folio, 122 maps and plates.
Cartes et descrip. gén. et particulières ... au sujet de la succession de la Couronne d'Espagne, 1701–2. 20 maps.
Petit et Nouveau Atlas, oblong 4to, 1705.
Les Postes de France et d'Italie, 1700, 1728, 1760.
Le Théâtre de la guerre dessus et aux environs du Rhein, 4to, Paris, 1705, with 26 maps.

De Fer also issued numerous maps of the various provinces of France, some being on several sheets, of most of which there were many editions.

An earlier worker in the century was Nicolas Tassin, who produced *Plans et Profilz de toutes les principales villes . . . de France*, 1634, with 100 maps; other editions in 1636 and 1638, and:

Cartes gén. des Royaumes et Prov. de . . . Allemagne, 1633, quarto.
Cartes gén. de toutes les provinces de France, 1634, in folio.
Cartes gén. de toutes les costes de France, 1634, oblong folio.
Descriptions de tous les Cantons, Villes, Bourgs des pays de Suisses, 1635.
Cartes Générales de la Géographie Royalle, 1655, with 84 maps.

The height of French influence was reached in the 18th century. Supreme in so many of the arts, France in this century took the lead from Holland, and became the centre of geographical science. Without reaching the prodigious output of the Dutch in the preceding century, her contribution to the advancement of cartography was of the highest importance. It was in France that scientific mapping from exact ground observation was commenced, and speculative cartography finally abandoned. Triangulation was started in 1744. Yet as in this great period France could not produce that which she did not adorn, mathematical exactness accompanied design, and great artists like Boucher, Cochin, Monnet, Choffard, Gravelot, Marillier and others embellished the works of her geographers, and so the 18th-century French maps were not only the most advanced, but also the most beautiful of their time. The weakness of the French school was in their predilection for theorising. They abhorred a vacuum, and so where no real knowledge was available they filled in the gaps with theoretical conceptions, so that their maps were in effect frequently more misleading than those of their Dutch predecessors.

In 1718 appeared the *Nouvelle édition du théâtre de la guerre en Italie* by the widow of Jean Baptiste Nolin senior, with 15 maps, an *Atlas Général* by his son, also Jean Baptiste, appearing in 1783 (44 maps and 4 plates).

The most prominent figure at the beginning of the century was Guillaume De l'Isle, born in 1675 in Paris, where he died 1726. Educated under the direction of his father, who was also a geographer, it is said that at the age of eight or nine years old he could draw maps to demonstrate ancient history, and he was regarded as an infant prodigy. In 1699 he produced a map of the world, maps of the four continents and two globes, terrestrial and celestial.

30 NICOLAS BEREY'S PLAN OF PARIS, 1645 (17" × 12")

31 MELCHIOR TAUERNIER: POST ROADS OF FRANCE, 1632 (20¾" × 16")

He was elected member of the academy in 1702 and in 1718 made Premier Géographe du Roi. De l'Isle instituted many reforms in his maps, and his work was highly rated, not only by his own countrymen, but by the world at large (69). His brother Nicolas became the friend of Peter the Great, and supplied him with information on the Russian Empire. His maps were copied or used as a basis by contemporary and later geographers, in fact, he had a lawsuit with Nolin, who was convicted of plagiarising. In all he issued over 100 maps of modern and ancient geography. His *Atlas Nouveau* was first issued by Covens and Mortier in Amsterdam, undated, but about 1708. This had 24 maps. Another edition appeared about 1730 with 54 maps; 1733 with 64 maps, and an even more extensive edition c. 1745, with 116 maps. The later editions of De l'Isle were published by Covens and Mortier in Amsterdam. Various copies do not agree as to the number of plates, the publishers either at the request of purchasers or arbitrarily varying the contents. A rare work is that of the discoveries of De Fonte 1753 with 4 maps, by Nicolas De l'Isle.

Claude De l'Isle, father of the former, was a geographer and historian, born in Lorraine in 1644, dying in Paris 1720. Besides many historical and genealogical works he published an *Introduction to Geography* in 1736. Guillaume De l'Isle had two brothers, Simon Claude, an historian who curiously enough was born the same year as his elder brother and died in the same year, and a younger brother, Joseph Nicolas, 1688–1768, who became a celebrated astronomer. He was in charge of the Royal Observatory in St. Petersburgh and stayed in Russia for twenty-two years, returning to France in 1747. His *Atlas Russicus* came out in 1745: it had 20 maps.

Philippe Buache, born 1700, died 1773, succeeded De l'Isle, issuing various atlases in the latter part of the century (94), namely, *Atlas Géographique de Quatres parties du Monde* (1769–99), *Atlas Géographique et Universelle* (1702–62) and *Cartes et Tables de la Géographie Physique* (1754), with 20 plates. Inferior in capacity to De l'Isle and D'Anville, he is principally known for his system of physical geography. He was one of the first to map the submarine world, his theory being to divide the world into a series of cavities or basins, both terrestrial and aquatic, the former based on the principal rivers of the world, the latter on a series of mountain chains on the sea bed, traceable according to Buache by the exposed points of islands. This led him to compose some of the most fantastic maps of the period. His nephew, Jean Nicolas Buache de la Neuville (1741–1825), was also a geographer, principally of marine maps.

One of the most celebrated of French geographers was Jean Baptiste Bourguignon d'Anville, born 1697, died 1782. He spent the whole of his life studying geography, and amassed a great number of maps, which were acquired by Louis XVI in 1779. d'Anville published his first map at the age of fifteen, this being a representation of ancient Greece. All his life d'Anville had a predilection for ancient geography, his *Géographie Ancienne et Abrégée* appearing in 1769, 1775 and 1810, and English translations in 1775, 1795, 1801, 1806, 1810 and 1820. d'Anville corrected the errors of Sanson and De l'Isle, issuing an *Atlas Général* without title from 1740 onwards with a varying number of maps. Another of his works was the *Nouvel Atlas de la Chine*, 1737 (with 42 maps).

Another family of geographical note was the Robert de Vaugondys (Gilles, 1688–1766, and Didier, 1723–86), who produced their *Atlas Universel* in 1757 (108 maps), the basic material for which came to the Vaugondys from Pierre Moulard Sanson, grandson of the founder of the family, Nicolas. The engraving for the *Atlas Universel* was entrusted to artists of ability, and charming cartouches were designed to adorn the maps. The court headed by Madame Pompadour supported the venture, there being over six hundred subscribers. It was priced at 126 livres, with 12 livres extra for binding in calf. R. J. Julien published his *Atlas Géographique et Militaire de la France* in 1751. These atlases merit a place in any collection, for they mark the transition from speculative cartography to exact observation on the ground.

The finest work of the century from the scientific point of view was, however, that of the Cassinis, who started triangulation in France in 1744, at first under official auspices and later at their own charge. After forty-five years' laborious work, the *Carte Géométrique de la France* was published in 182 sheets. Louis Capitaine produced a reduction of this magnificent work in 1789 on 24 sheets.

Two other cartographers deserve separate mention: Bellin and Le Rouge. Jean Nicolas Bellin, born in Paris 1703, died at Versailles in 1772. Extremely industrious, he was attached to the French Marine Office. His works are as follows:

Atlas Maritime, 1751.
Neptune Français, 1753.
Mémoires sur les Cartes des Côtes de l'Amérique, September 1755.
Hydrographie Française, 1756–65, folio, 2 vols. (86 maps) (the number of maps varies, another
    copy 1737–76, with 105 maps, and another 3 vols. with 149 maps).
Guyane, 1757.
Description Géographique des Antilles, 1758.
Essai Géog. sur les Iles Britanniques, 1763, 4to.
Petit Atlas François, 1763, 4to, 5 vols. (581 maps). Reissued as *Petit Atlas Maritime*, 1764.
L'Enfant Géographe, 1769.
Corsica, 1769.
Gulf of Venice and Morea, 1772.
Débouquemens . . . de l'Ile de St. Domingue, 1773 (24 maps).

G. L. Le Rouge specialised more on atlases of plans and fortifications. He was educated as a military engineer. His works are:

Théâtre de la Guerre en Allemagne, 1741, 4to (65 plates).
Nouvel Atlas Portatif, 1748 (87 maps).
—— another edition, 1756, 2 vols., 4to (192 maps).
Description du Château de Chambord, 1750, folio.
Recueil des Plans de l'Amérique, September 1755, 8vo (23 plans).
Recueil des Côtes Maritimes de France, 1757, 4to.
Atlas Prussien, 1758, folio, 25 ll.
Recueil des Villes, Ports, d'Angleterre, 1759, 8vo (17 maps).
Topographie des Chemins de l'Angleterre, 1760, 8vo (101 maps).
Recueil des Fortifications, Forts, et Ports de mer de France (1760), 8vo (89 maps).
Curiosités de Londres, 1765.
Curiosités de Paris, 1768.
Atlas Amériquain Septentrional, 1778, folio (20 maps).
Pilote Amériquain Septentrional, 3 parts, 1778–9, folio (20, 21 and 13 maps).

The Library of Congress possesses in addition a collection of 77 folio maps without title, probably separate maps issued by Le Rouge, bound together. They date from 1740 to 1747. Le Rouge's works on England and America were not original, but reductions or copies of English surveys, and after Bellin. The last two works are rare.

A very decorative marine atlas was issued by Louis Renard in 1715. This contains 28 charts with large ornamental title-pieces in the Dutch style (32). This was issued in three ways: uncoloured, the body of the map coloured and the cartouches left plain, and fully coloured. The last examples are very fine.

Other publications include atlases by Bonne (*Atlas Portatif*, c. 1785 and *Atlas Encyclopédique*, 1787); Bougard (*Petit Flambeau de la Mer*, 4to, 1789, first issued in 1684; English edition, 1801); Desnos (*Atlas Général Méthodique*, 1768), and Janvier, who issued 7 maps of the world, the four continents, France and Germany in 1760–61, reprinted on a reduced scale in 1762 in the *Atlas Moderne*, in 1763 in Gourmet's atlas and in 1776 by Santini.

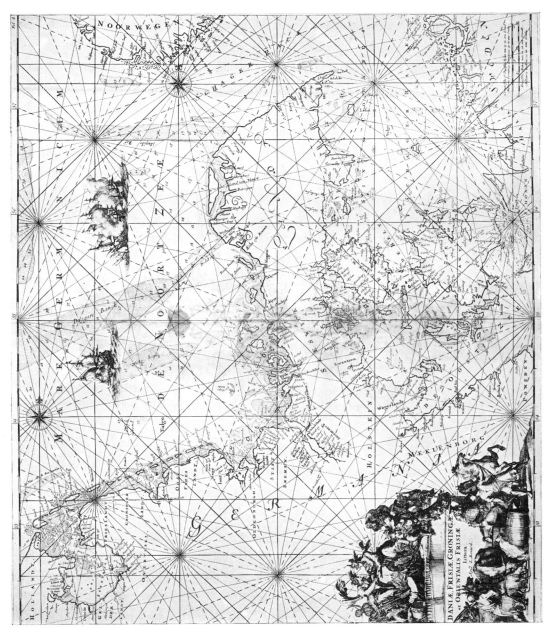

32 L. RENARD'S CHART OF THE BALTIC, ETC. (22″ × 19½″)

from *Atlas de la Navigation et du Commerce*, Amsterdam, 1715

33 EARLY SEVENTEENTH-CENTURY
TITLE-PIECE

34 MID-SEVENTEENTH-
CENTURY DEDICATION

35 LATE SEVENTEENTH-CENTURY TITLE-PIECE
SEVENTEENTH-CENTURY DUTCH CARTOUCHES

Plans of Paris have a long and more or less continuous history from the 16th century onward, some of the principal being as follows: 1540, Münster; 1551, Truschet and Hoyau; 1572, Braun and Hogenberg; 1609, Nicolay; 1615, Merian; 1676, Bullet and Blondel; 1675, Jouvin de Rochefort; 1697, De Fer; 1739, Bretez (twenty sheets, usually bound in calf with the arms of the city on the sides); and 1754, Rocque on eight sheets.

At the end of the 18th century the internal troubles of France helped to end her supremacy, and general initiative and primacy in map production passed from France to England, then the main seat of wealth and maritime power.

AUTHORITIES

ADVIELLE (V.). *Description du Berry et Diocèse de Bourges*, Paris, 1865.

ANTHIAUME (A.). *Cartes Marines . . . Voyages et Découvertes chez les Normands, 1500–1650*, Paris, 1916.

—— "Un pilote et cartographe Havrais du XVI Siècle : Guillaume Le Testue" (*Bull. de Géogr. Hist. et descript.*, Paris, 1911).

BAGROW (L.). *A. Ortelii Catalogus Cartographorum*, 1928–30.

BERTHAUT (Col.). *Les Ingénieurs Géographes Militaires 1624–1831*, 1902.

BOSSEBOEUF (L'Abbé). *La Touraine et les Travaux de géographie*, Tours, 1894.

BROWN (Lloyd A.). *J. D. Cassini and his world map of 1696*, Ann Arbor, 1941.

Catalogue des Cartes, Plans et vues de côtes qui composent l'Hydrographie Française, Paris, 1832.

D'AVEZAC (M.). *Coup d'œil historique sur la projection des cartes de Géographie*, Paris, 1863.

DENUCÉ (J.). "Jean et Jacques Surhon cartographes montois d'après les archives Plantiniennes" (*Ann. du cercle Arch. de Mons*, 1914).

DOUBLET (E.). "Une famille d'Astronomes et de Géographes (De Lisle)" (*Revue Géogr. Comm.*, 1934).

DRAPEYRON (L.). "Le premier atlas national de la France, 1589–94" (*Bull. de Géogr. Hist. et descript.*, Paris, 1890).

—— *L'évolution de notre premier atlas national sous Louis XIII*, Paris, 1890.

—— *Notre premier atlas national et la Ménippée de Tours sous Henry IV*, Paris, 1894.

FERRAND (H.). "Les premières cartes de la Savoie" (*Bull. de Géogr. Hist. et descript.*, 1906).

—— "Les premières cartes du Dauphiné" (*Ann. du Club Alpin français*, 1903).

FORDHAM (Sir H. G.). *An Itinerary of the 16th Cent. La Guide des Chemins d'Angleterre, Jean Bernard, Paris, 1579*. Cambridge, 1910.

—— "Liste alphabétique des Plans de Villes, Citadelles et Forteresses qui se trouvent dans le grand atlas de Mortier édn. d'Amsterdam de 1696" (*Bull. de Géogr. Hist. et descript.*, 1911).

—— "The Cartography of the Provinces of France, 1570–1757" (*Studies in Carto-Bibliography*, Oxford, 1914).

—— *Note on a series of early French atlases, 1594–1637*, London, 1921.

—— *Les Guides-Routiers, Itinéraires et Cartes Routières de l'Europe, 1500–1850*, Lille, 1926.

—— *Une carte routière de France du XVII siècle*. La Géographie, Paris, 1926.

—— *Maps, their History, Characteristics and Uses*, Cambridge, 1927.

—— *Les Routes de France, Etude Bibliographique sur les cartes Routières, etc.*, Paris, 1929.

French Maps of the World, drawn in 1536, 1546 and 1550, by C. H. Coote. 3 portfolios and 4to text. Privately Printed, 1898.

GAFFAREL (P.). "André Thevet" (*Bull. Géogr. Hist. et descript.*, Paris, 1888).

GALLOIS (L.). *De Orontio Finaeo Gallico Geographo*, Paris, 1890.

—— "Les origines de la carte de France: La carte d'Oronce Finé" (*Bull. Géogr. Hist. et descript.*, 1891).

—— "Waldseemüller chanoine de St. Dié" (*Bull. Soc. Géogr. de l'Est*, Nancy, 1900).

—— "La Grande Carte de France d'Oronce Finé" (*Ann. de Géogr.*, 1935).

GRAVIER (G.). *Les Normands sur la Route des Indes*, Rouen, 1880.

HELBIG (H.). "Gilles de Bouillon, sa vie et ses ouvrages" (*Bib. Belge*, 1889).

JOMARD (E. F.). *Les Monuments de la Géographie*, Paris, 1842–62.

LANGLOIS (L.). "L'Atlas de Bouguereau" (*Bull. Soc. Arch. de Touraine*, 1902).

LE PARQUIER (E.). "Note sur la carte générale du pays de Normandie" (*Soc. normande de Géographie*, Rouen, 1900).

L'Isle (G. De). *Liste des Ouvrages de*, Paris (1733).

Marcel (G.). "Louis Boulengier d'Alby astronome, géomètre et géographe" (*Bull. de Géogr. Hist. et descript.*, Paris, 1890).

—— *Notes sur quelques acquisitions récentes de la Section des cartes et collections géographiques de la Bibliothèque Nationale*, 1897.

—— "A propos de la carte des chasses" (*Revue de Géogr.*, Paris, 1897).

—— "Les origines de la carte d'Espagne" (*Revue hispanique VI*, Paris, 1899).

—— *Un almanach xylographique à l'usage des Marins Bretons*, Paris, 1900.

—— *Le Plan de Bâle et Olivier Truchet*, Paris, 1902.

—— "Une carte de Picardie inconnue et le géographe Jean Jolivet" (*Bull. de Géogr. Hist. et descript.*, Paris, 1902).

Morren (E.). "Charles de l'Escluse, sa vie et ses Œuvres" (*Bull. de la Fédération de Soc. d'horticulture de Belge*, 1874).

Nordenskiöld (A. E.). *Facsimile Atlas*, 1889.

Nouvelle Biographie Générale, Paris, 46 vols., 1857–66.

Phillips (Philip Lee). *List of Geographical Atlases in the Library of Congress*, 4 vols., Washington, 1909–20.

Public Archives of Canada National Map Collection. *French Atlases in the Rare Atlas Collection*, 1974.

Rodger (E.). "An Eighteenth-Century Collection of Maps connected with Philippe Buache" (*Bodleian Library Record*, Vol. VII, 1963).

Roland (F.). *Les cartes anciennes de la Franche Comté*, Besançon, 1913.

Sandler (C.). *Die Reformation der Kartographie um 1700*, München und Berlin, 1905.

Smet (A. de). *Le général-comte de Ferraris et la carte des Pays-Bas autrichiens*, Industrie Bruxelles, 1966.

Tooley (R.V.). *Map Making: France XVI-XVIII Centuries*, reprint Huguenot Society, London, 1952.

—— *French Mapping of the Americas; the De l'Isle Succession*, Map Collectors' Circle No. 33, 1967.

—— *Maps of South West France*, Map Collectors' Circle, 1966.

Vacher (A.). "La Carte de Berry par Jean Jolivet 1545" (*Bull. de Géogr. Hist. et descript.*, Paris, 1907).

Vallée (L.). *Catalogue des Plans de Paris et des Cartes de l'Ile de France . . . conservé . . . Bib. Nat.*, Paris, 1908.

Vaugondy (R. de). *Essai sur l'histoire de la Géographie*, Paris, 1755.

Vayssière (A.). *Générale Description du Bourbonnois par Nicolas de Nicolay*, 1889.

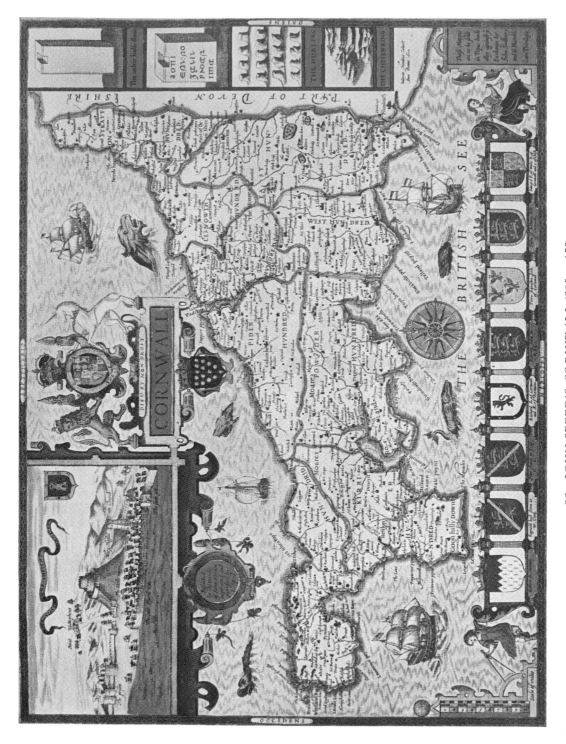

36 JOHN SPEED'S *CORNWALL* (20″ × 15″)
from *The Theatre of the Empire of Great Britain*, 1611

CHAPTER VII

English Map-makers

IN the history of map-making, England has a brilliant but somewhat unequal history. Map-making of the past on a large scale has always followed sea power and wealth, and the loss of either to a consequent decline. Thus Italy was the first in modern times to take the lead in all matters relating to map production when she was at the height of her commercial prosperity. The maritime successes and trade of the Dutch gave them predominance from the late 16th to mid-17th centuries, and the consolidation of power and expansion of France under Louis XIV transferred that pre-eminence to France. From the end of the 18th and during the 19th centuries England was the premier maritime and commercial power, and made the greatest contribution to cartography. The Carys, the Arrowsmiths, William Smith, and the Ordnance (94) and Admiralty surveys secured the scientific advancement of Britain. In Christopher Saxton, England had the pioneer of national atlases; in Norden, the inventor of the triangular distance tables; in Ogilby, the founder of the road map to the whole country; in Halley, the originator of the meteorological and magnetic chart; in William Smith, the pioneer of the geological map. Even in the heyday of Dutch map production, a not insignificant part of their world mapping was based on English surveys (the work of Drake, Cavendish, Ralegh, Frobisher, Hudson, Smith and others). At a later date Cook's work was universally copied, and India, Africa, Australia, the Arctic and parts of America owe their mapping largely to British endeavour. In the large-scale county maps of the 18th and early 19th centuries Britain has a cartographical development without parallel elsewhere. Greenwich was adopted as the prime meridian in the 1911 conference for the international map of the world. Finally, in the British Museum, Britain has the largest and most valuable collection of ancient cartographic monuments, freely available to the students of all nations.

Compared with many lands, England has a comparative wealth of early cartographic material. The oldest representation of Great Britain, as with most other countries of Europe, is that detailed by Ptolemy about A.D. 150, first printed in Bologna in 1477 (for later editions see Chapter I). This map, with its rectangular diminishing squares, presents a very handsome appearance, the sea being fully engraved with waved lines, white scrolls being left in places to name the seas, the land left white, with the few names standing out boldly and clearly.

The oldest extant native map is the so-called Anglo-Saxon map formerly in the possession of Battle Abbey, now in the British Museum. It dates from the end of the 10th century, and is a world map with the east at the top. Though inferior to Ptolemy, with Asia shown at right angles to the coast of North Africa, it is nevertheless recognisable and far superior to much later mediæval work. It is also the first map so far as is known to add to the knowledge of Ptolemy as regards North-west Europe. The British Isles are prominently marked, with

Ireland running east and west instead of north and south. Scotland curiously enough is twisted to the left instead of the right as in Ptolemy. Jerusalem is not in the centre, as in most mediæval maps.

37 *Anglo-Saxon Map, 10th Century*

The next historical records of importance are the maps compiled by, or for, Matthew Paris in the scriptorium of the Benedictine Abbey of St. Albans about 1250. There are four extant MSS. of Matthew Paris's *Chronica Maiora* and his *Historia Anglorum* and John of Wallingford's chronicle (died 1258), with accompanying maps. They are the earliest-known detailed maps of Great Britain. They differ in detail and even outline, and have a varying number of place-names, 84 in the smallest imperfect example, and 280 in the fullest example. Four of the maps were excellently reproduced by the British Museum in 1928.

John of Holywood or Halifax, better known under the latinised form of his name Johannes de Sacrobosco, was probably a Yorkshireman. In 1230 he migrated to Paris and compiled his *Tractus de Sphaera* (a manual of astronomy and cosmography accompanied by a world map). This treatise had an enormous vogue in the Middle Ages, twenty-four editions appearing before 1500 and a further forty editions up to 1647. It was translated into Italian, French, German and Spanish, and its commentators included men of the calibre of Melanchthon, Regiomontanus, Pierre d'Ailly, and Clavus. The map is of the hemispherical climatic type, with a rough delineation of the known world. The east is at the top. Though containing no original work, being based on Macrobius, it was nevertheless the map that had the greatest circulation among students from the 13th to the 15th centuries.

About 1280 the Hereford Mappa Mundi was composed by Richard of Haldingham, prebend of Lincoln, Hereford and Salisbury Cathedrals (9). It is one of the finest extant mediæval world maps, circular in shape, the sea green, the rivers blue, the Red Sea red, covered with little miniatures of notable cities and mythological monsters and people. England, Scotland and Ireland, shown as separate islands, are configured to suit the rotundity of the map.

In many mediæval world maps the British Isles are shown diagrammatically, or merely named either within or without a lozenge in the encircling ocean.

In the next century, about 1335, a remarkable map of Great Britain appeared. Its author is unknown, and so it is called the "Gough map" (13), after Richard Gough its discoverer, or the "Bodleian map," after its resting-place in Oxford. It shows a great advance on Paris's maps. England is remarkably well mapped, Wales slightly less so, and Scotland in rudimentary form. Roads, or rather mileage, are shown between main towns by means of straight connecting lines. The symbols employed are most interesting, walled and unwalled towns being differentiated, ecclesiastical sites marked, forests named, etc. The roads are coloured red. Many towns, especially in Wales, are shown unnamed. It was reproduced by Gough and by the Ordnance Survey. It is a map that richly deserves a special monograph of its own.

About this time (1340 onwards) the estate map began to make its appearance. Many examples, mainly of later periods, have been preserved, some of which have been described by Fordham, Lynam and the Essex County Council. An interesting and valuable relic of this period is a mediæval map of Sherwood Forest drawn about 1376, belonging to the Duke of Rutland.

In 1489 a world map was issued with Johannes Eschuid's (or Eastwood's) work—*Summa Anglicana*. This was only a reversed copy of Macrobius.

In the beginning of the 16th century two small maps of the British Isles appeared, the first by Pietro Coppo was a wood-cut engraved about 1525. Its outline was based on the portolans, its interior nomenclature on Ptolemy, with a few modern additions. The other was a map by Benedette Bordone in his *Isolario* of 1528, re-issued in 1533 and 1534.

In 1540 a separate map of England appeared in Munster's edition of Ptolemy, possibly based on the Gough model, but less detailed. It is a woodcut, and the first separately printed map of England.

In 1546 appeared the first copper-plate map of the British Isles based on contemporary knowlege. It is ascribed to George Lily, a Catholic refugee in the service of Cardinal Pole. A reproduction of this fine map was issued by the British Museum in 1928 and by George Beans in 1934, with descriptive text by Edward Lynam. The editions of Lily's map are as follows:

1546	Rome. G.L.A. (Georgius Lilius Anglus).	1555	London. T. Gemini.
1549	Antwerp. Joannem Mollijns. Woodcut.	1556	(Rome.) Signed I. H. S.

1556	Venice. Vavassor. Woodcut.	1563	Venice. Camocius.
1558	Rome. Sebastianus a Regibus Clodiensis.	1589	Rome. Marc. Clod. typis.
1562	Venice. Bertelli.		

Of the first edition, at least seven copies are known; of the editions of 1549 and 1555, only single examples (both in the Bibliothèque Nationale, Paris), and the remainder from several examples, though they rarely come on the market. These derivations of Lily's map are not identical either as to size or information, two are wood-blocks, the rest copper-plates. The first three, and the Rome editions of 1558 and 1589, have the north on the right of the map.

Lily's map was superseded in 1564 by the large map of the British Isles compiled by Mercator. This magnificent map was printed on eight copper-plates on a scale of $14\frac{1}{2}$ miles to the inch, with the title "Angliae Scotiae & Hiberniae noua descriptio," 1564. The source from which Mercator obtained his data for this remarkable achievement is unknown, though it must obviously have been English in origin. It is known that fourteen copies of this map were dispatched to England by Plantin, but the only recorded examples to date are preserved in Breslau, Rome and Perugia. A reproduction was issued in Berlin in 1891.

A year earlier Laurence Nowell, Dean of Lichfield, compiled a map covering the British Isles on nineteen sheets, three being devoted to Scotland, three to Ireland and thirteen of various English counties, mainly together in groups. Degrees of latitude and longitude are marked, and in places, particularly in northern England, Anglo-Saxon names are included. The scale is naturally small, and it was never printed.

In 1573 Ortelius, in his *Additamentum* (and in subsequent editions of his atlas), printed separate maps of England and of Wales, both by Humphrey Lhuyd, the first really good detailed maps of both these kingdoms. Both are very decorative, fairly easily obtainable and not expensive.

The finest map of the 16th century was, however, Saxton's large map of England and Wales, issued in 1583 with the title "Britannia Insularum in Oceano Maxima," engraved on twenty-one sheets, on a scale of approximately 8 miles to the inch. It was known only from a late reproduction up to 1930, when Messrs. Francis Edwards discovered an original issue. This was fortunately acquired by the British Museum. In 1938 the same firm found another even finer example of this map, differing from the former in that it had a broad engraved border containing eighty-five coats-of-arms of the nobility and gentry. This is now in the Birmingham Public Library.

Philip Lea secured the copper-plates, printed a fresh title, and considerably altered the plates, changing the coats-of-arms of the nobility and wording and adding the principal roads which he took from Ogilby's survey, 1675. He also changed the Elizabethan ships to a Carolinian aspect. This re-issue appeared in 1687, engraved by Sutton Nicholls. It is rare: the British Museum has a copy in proof state before the alterations were complete. It was also issued by T. Bowles and John Bowles and Son, and again by Robert Sayer in 1763.

Saxton's map was also copied on a reduced but still large scale in 1644. This map is known as the Quartermaster's Map. It was engraved by Hollar and published by Jenner. It was printed on six sheets, folded to fit into the pocket, with a title, "The Kingdome of England & Principality of Wales exactly described . . . in six maps, portable for every man's pocket." This production was re-issued by Jenner in 1671 with a few roads inserted and by John Garrett (1688), by John Rocque in 1752, and again in 1799. Thus Saxton's map had an effective life of over two hundred years.

Another large-scale map based on Saxton was that by John Adams, "Angliae totius Tabula cum distantiis notioribus in itinerantium usum accommodata," on twelve sheets, dedicated to William III by Philip Lea (1692), the distances between towns and villages being shown in straight lines. The prime meridian is based on London. Saxton's surveys were also used by other map-makers, both in general maps of the kingdom and in county maps.

38   CHRISTOPHER SAXTON'S MAP OF DORSET (21″ × 15″)

from his *Atlas* 1791

In 1588 Robert Adams drew a series of charts showing the engagements between the English Fleet and the Spanish Armada (42), including the adjoining coast, and these were beautifully engraved by Augustine Ryther about 1590 on ten sheets, with a general map and a title "Expeditionis Hispanorum in Angliam vera descriptio."

Another rare map of the period is Plancius's map of the British Isles of 1592, engraved by Baptista Doeticum. It measures 55 by 39 cm. England and Wales are based on Saxton, but with additions round the coast. Only two copies are known to date. This map was re-engraved on a slightly smaller scale in 1594 with different ornamentation, a portrait of Queen Elizabeth being introduced. Finally, it was re-issued in the 1606 edition of Ortelius, but this time without the portrait of the Queen. In 1603 a map of the British Isles was engraved by William Kip for John Woutneel. This map shows more names than the 1594 Hondius map on which it is based. A large map, it measures 81 by 103 cm., Scotland being copied from Ortelius and Ireland from Boazio, though with considerable variations from the usual form, the only original feature being the introduction of small tents to mark the sites of ancient battlefields.

The first English incursions into the wider fields of foreign cartography began in the 16th century: e.g., John Rotz, geographer to Henry VIII, compiled a world map in 1542. Sebastian Cabot's map of 1544 has been lost, but Borough's chart of the North Atlantic (1576), said to have been used by Frobisher, has survived. Anthony Jenkinson, a member of the Mercers Company, was appointed leader of an expedition sent to Russia in 1557 to open up the eastern trade routes for the Muscovy Company. An intelligent and observant man, Jenkinson compiled an excellent map of the country, his map being used by Ortelius in his *Theatrum*. Jenkinson and Humphrey Lhuyd, who composed the maps of England and Wales, are the only two Britons whose work was used by Ortelius. Early English navigators, such as Chancellor, Drake, Gilbert, Frobisher, Davis and Ralegh, naturally composed charts and maps of their discoveries, and these caused lively interest and discussion among the learned in England; for example, the celebrated Dr. Dee, who took a particular interest in the North-West Passage, and Burleigh, Elizabeth's counsellor. Few, however, were printed, except indirectly through foreign sources. There are manuscript maps in the British Museum formerly in the possession of Dr. Dee, dating from 1580 and 1582, showing North America and the Arctic regions. Michael Lok was one of the early surveyors of these regions, and his map was used by Hakluyt in 1582, as also was Robert Thorne's map of the world. Both these maps were printed from wood-blocks. Another world map was that of George Beste in 1578. He likewise drew a map of Frobisher's Strait. A manuscript map of Guiana by Ralegh is preserved in the British Museum. It dates from about 1596. Hondius, the Dutch publisher, used Ralegh's information in the construction of his printed map in 1599. An even more famous map, likewise showing English effort, was the world map delineating the voyages of Drake and Cavendish. This was published in Antwerp about 1581 and again about 1595 published by Hondius. Both these maps were reproduced by the British Museum in 1931. In 1592 Emery Molineux constructed a terrestrial globe, and in 1596 John Blagrave, an Oxford mathematician, compiled a world map on an unusual projection. This was engraved by Benjamin Wright.

A much finer map of the world was compiled by Edward Wright in 1600 for *Hakluyt's Voyages*. It is one of the treasures of cartographical collections, rarely being found with the book. Edward Wright, born about 1558, died 1615. He was educated at Cambridge, and served in the expedition to the Azores under the Earl of Cumberland. His great work, *Certain Errors of Navigation*, was completed in 1592, but not published till 1599. It effected a complete revolution in nautical science. Rather rashly Wright gave a copy of his manuscript to the Dutch publisher J. Hondius to peruse, with the result that Hondius utilised Wright's projection in the maps published by him in 1599 (map of the world and the four continents) without due acknowledgment. Wright's own version of his world map appeared in 1600 in

Hakluyt's Voyages. Another version appeared in 1610. Wright also assisted Gilbert in the compilation of his work on the magnet (1600), and translated Napier's Logarithms (1614) into English. Wright's work was extensively copied and utilised in Holland, and practically speaking the so-called Mercator charts in use at the present time are drawn on the projection laid down by Wright.

The English edition of Linschoten, printed by John Wolfe in 1598, has 12 folding maps copied from those of the original Dutch edition.

The 17th century witnessed a considerable advance in English cartography. In 1606, an English translation of Ortelius, the first general atlas of the world to be published in English; in 1612, Captain John Smith's map of Virginia appeared engraved by William Hole, of which there are several states; and in 1624, Smith's *General History of Virginia*, re-issued in 1627: this work had four maps—Ould Virginia, Virginia, the Summer Islands and New England.

In 1619 William Baffin compiled a map of the Mogul Empire, the best up to that date, from information supplied by Sir Thomas Roe. It was engraved by Elstrack. Various manuscript plans by English pilots have survived from this period; for example, a plan of Algiers in 1620 by Richard Norton, master gunner and engineer.

In 1627 John Speed published his *Prospect of the Most Famous Parts of the World*, the first printed general atlas by an Englishman. It was combined with *The Theatre* with county maps of the same date, and contained 22 maps. It was re-issued in 1631, 1646 and 1662, and in 1676 it was again reprinted with the addition of new maps, viz.:

Virginia and Maryland.
New York and New England.
Jamaica and Barbados.
Carolina.
East Indies.
Russia.
Palestine.

Miniature editions appeared in 1627, 1646, 1668 and 1675.

In 1635 Michael Sparke and Samuel Cartwright printed *Historia Mundi, or Mercator's Atlas, with new maps and tables by the studious industry of Judocus Hondy, Englished by W S(altonstall)*. This small folio is of special interest as containing R. Hall's map of Virginia and Smith's of New England. The Virginia map is usually missing, as it was not actually issued till a year later, but was supplied to purchasers of the original work on application. This first edition is rare. A second edition was published in 1637. Even this second edition often lacks Hall's and Smith's maps. The remainder of the maps are printed from copperplates inset into the text.

In 1633–6 an English edition of Mercator appeared under the title *Atlas or Geographic Description of the Regions, Countries and Kingdoms of the World ... trans. by Henry Hexham, Quartermaster to the Regiment of Colonel Goring. Printed Amsterdam by Henry Hondius and John Johnston* [i.e., Jan Jansson], in 2 vols., folio, with 196 maps.

In 1646 an important atlas appeared—the *Arcano del Mare* of Sir Robert Dudley, the first marine atlas in which all the charts were drawn on Mercator's projection. Though printed in Italy, this is the first English sea atlas. Dudley was a skilled mathematician and navigator, and two of his friends were Kendal and Davis, both expert mariners. He made a voyage to the West Indies, the Orinoco and Guiana in the *Earwig* and the *Bear*. Turning Roman Catholic, he emigrated and settled in Florence, where his great work was produced. The engraver Lucini took eight years to engrave the copper-plates. A second edition appeared in 1661 (91).

John Seller, Hydrographer to Charles II and James II, had a prolific output of maps, charts and geographical and nautical publications, and was granted a monopoly for the former for thirty years. Apart from his marine works—to be mentioned later, he issued:

39 RICHARD BLOME'S *CHESHIRE* ($9\frac{1}{2}'' \times 7\frac{1}{4}''$)
from *Geographical Descriptions of the World*, 1693

40 MICHAEL DRAYTON'S "RUTLAND" (12″ × 9¾″)
from his *Poly-Olbion*, 1612–22

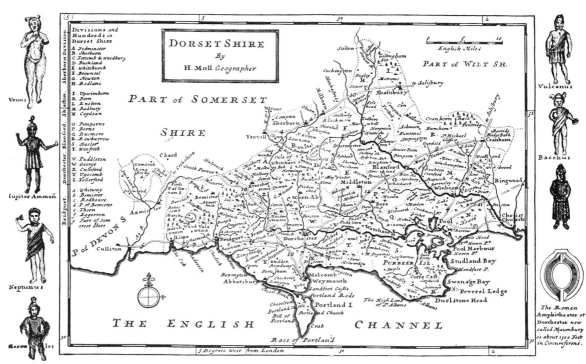

41 HERMAN MOLL'S *DORSETSHIRE* (10¼″ × 7¼″)
from *A New Description of England and Wales*, 1724

Atlas Minimus, 1679 (53 maps),
New System of Geography, 1685 (63 maps),
—— 1709 (71 maps),

Anglia Contracta, 1695, 8vo (62 maps),
Atlas Terrestris, 1680–85, folio;

and large-scale maps of a few counties. The *Atlas Terrestris* varies as to its contents, no two copies I have examined agreeing as to the number or identity of the maps. The B.M. copy has 16 maps, and I have seen other examples with 24 and 34 maps. Seller took the majority of his foreign maps from the Dutch publishers De Wit, De Ram, Danckerts, Allard and Visscher, etc., sometimes erasing the Dutch title and substituting an English one, or, if Latin, leaving them untouched. He did, however, include a certain number of English productions, including maps by himself, of Tobago (86), Jamaica, and the XVII Provinces (Netherlands), maps and plans of Tripoli and the Island of Sicily by W. Hollar, maps by Overton and Morden, and by Philip Lea of the West Indies, Carolina, Jamaica, New England, Savoy and Hungary.

Despite Seller's monopoly, William Berry, a contemporary, issued a series of maps of large folio size based on the maps of Sanson, and he is sometimes in consequence called the English Sanson. These maps were issued and sold separately, but about 1689 Berry collected them together as an atlas, though without a title. The example in the British Museum has 37 maps and 7 geographical tables and the Library of Congress (38 maps). The Greenwich Museum copy contains 38 maps. Messrs. Maggs had a copy containing 69 maps, the extra maps including town plans and maps of the English plantations in America by Morden, Moll and Lea.

Another map publisher of the period was Moses Pitt, whose *English Atlas*, 4 vols., 1680–83, was issued at Oxford. Pitt planned a large 12-volume world atlas in English in the Blaeu-Jansson tradition, but his ambitious effort failed; only four volumes appeared, Pitt was ruined and imprisoned in the Fleet for debt. Pitt's maps had neither the quality nor the delicacy of the maps being drawn in Holland, though they were produced in a handsome format, some copies being done on large and thick paper, red ruled and sumptuously bound.

A popular work was Peter Heylin's *Cosmography*, issued in small folio in 1652, re-issued 1657, 1660, 1666, 1669, 1670, 1674, 1677, 1682 and 1703. This was illustrated with 4 folding maps of the four continents. A second work was Blome's *Cosmography*, 1682, with a world map, maps of the four continents and individual countries. They were mostly engraved by Francis Lamb, but one was by Hollar and one by Thomas Burnford. It was re-issued in 1693. Blome has been roughly handled by the critics, accused of plagiarism and downright theft. It is true he did no original work and that much of the engraving done for him is not in the best manner, yet he is far from being the only sinner in either respect; and if he does not please the purist or specialist, at least his maps are typical of his period, not without a certain quaint charm, and have given much pleasure to the many collectors whose prime consideration is not the utility of their maps, but their decorative quality or the period atmosphere they evoke.

A fairly common atlas printed at the end of the century is that by Edgar Wells, a *New Set of Maps of Ancient and Present Geography*, printed at Oxford about 1700, each map being dedicated to Humphrey, Duke of Gloucester. A second edition appeared in 1704. The maps, though crudely drawn, are not without a degree of decorative attraction, though their geographical value is strictly limited.

The century, which opened brilliantly with Edward Wright, closes with three further names famous in cartography, two of national, the other of international, repute. John Ogilby, born 1600 near Edinburgh, died 1676, had an extremely versatile career, being a bookseller, translator, printer, dancing master, theatre owner, master of the King's Revels, and finally a geographer. His claim to fame rests most securely on his last accomplishment. He made the first survey of the roads of England and Wales, the result being published in a folio volume entitled *Britannia, Vol. I . . . a Geographical and Historical Description of the Principal Roads thereof*, 1675. This had 100 copper-plates showing the roads in continuous strip form, six or seven to a plate, with a decorative heading enclosing the title (53). Ogilby was the first to adopt the statute mile of 1,760 yards. Before Ogilby different scales were in use in different parts of the country, for example short, middle and long English miles, but they can rarely be worked out accurately. Not only did he start a numerous progeny by his work—his *Britannia* being followed by a whole family of Road Books—but he revolutionised the printed map: from his time roads began to be inserted on the county maps, and this soon became an established practice. There were two issues of the *Britannia* in 1675, the first issue having 7 leaves of text on London, the second issue 4 leaves only. Some copies were done on large and thick paper red ruled. It was re-issued in 1698, the frontispiece by Hollar being discarded. The plates were reduced in size, a series of county maps added and other information, the whole being issued under a new title—*Britannia Depicta or Ogilby Improved*, by John Owen and Eman. Bowen, in 8vo size in 1720; re-issued 1721, 1723, 1724, 1730, 1731, 1734, 1736, 1749, 1751, 1755, 1759 and 1764. Ogilby was appointed to survey the city of London in conjunction with Morgan, Oliver and Mills, after the Great Fire, the result being a large-scale plan of London, 9 by 6 feet, published in 1677. Ogilby also issued historical and geographical accounts of Africa (1670), America (1670) and Asia (1673). These works, hand-somely printed in folio size, each have numerous engraved maps and plates, well engraved and extremely decorative. Ogilby was succeeded as Cosmographer Royal by a relative (his wife's grandson), William Morgan.

Little is known of Capt. Greenvile Collins, who flourished 1669–93, except that he served in the Navy and, when a younger brother of Trinity House, was appointed in 1681 to survey the coasts of Great Britain. This survey took him seven years, the charts he compiled being printed and issued as completed, and they were first issued collectively under the title *Great Britain's Coasting Pilot* in 1693 (48 charts on 45 sheets), re-issued 1723, 1738, 1749, 1753, 1756, 1760, 1764 and 1785. This work forms a landmark in the charting of Great Britain, being far in advance of any previous effort in this field. It is the first complete Pilot Book in English of the whole of the coasts of Great Britain, including the Orkney, Shetland and Scilly Islands, in some ways the equivalent for marine maps of Saxton's land maps. Some copies were done on large and thick paper coloured and heightened with gold. The chart of the Harwich and Woodbridge area is dedicated to Samuel Pepys, Fowey to the Bishop of Bristol, Falmouth to Sir Peter Killigrew, Islands of Scilly to the Duke of Grafton, Mouth of the Dee and Carreck Fergus to William III, the Thames and its Estuary to the Trinity Brethren, and others to simple ships' captains. Other charts include Rye, Dartmouth, Avon (44), Plymouth, Isle of Man, Leith, Aberdeen, Dublin and Belfast. The charts were engraved by J. Harris, F. Lamb and Js. Clark, James Moxon and H. Moll. Among the subscribers to the work were Charles II, James II, William III, Prince George, John Flamsteed, and the cities of Chester, Aberdeen, Edinburgh, Harwich, Liverpool, Newcastle and Yarmouth.

The other name to add a lustre to English cartography was that of Edmund Halley. Born in 1656 and dying in 1742, he was the inventor of the meteorological and magnetic chart. Halley is thus "one of the fathers of physical geography, the originator of graphical methods of representing on maps the geographical distribution of physical features of the earth." Halley's name will always be associated with the comet that is called after him. But astronomy was not the limit of his activities: his contribution to geography was momentous. At the age of twenty he spent two years in St. Helena, taking observations on the southern stars. He was a Fellow of the Royal Society, Professor of Geometry at Oxford 1703, and Astronomer Royal at Greenwich 1721. His contributions to cartography were:

(1) A Meteorological chart, 1688.

The first meteorological chart to be compiled, issued in the *Philosophical Transactions of the Royal Society*, No. 169. This chart was often reproduced in England and Holland and in books on navigation.

(2) The Magnetic chart, 1701: "Tabula haec Hydrographica Variationum Magneticarum Index."

The first edition, dedicated to William III, shows the Atlantic only. The second edition, dedicated to Prince George of Denmark, was extended to include the whole world (1702).

These are among the rarest of cartographical treasures.

From 1698 to 1700 he sailed in the *Paramour*, a pink, as commander, making observations in the North and South Atlantic, the first sea voyage undertaken for a purely scientific objec , and the track of his voyage is frequently shown on later maps. Halley's chart was a great success, and was frequently used and copied both at home and abroad.

To sum up the century, English contribution to the science of cartography, if small in quantity, was high in quality.

The 18th century witnessed a further advance. The Dutch school had declined towards the end of the previous century, and the rising French school never took so complete a hold as had the Dutch in its heyday, and towards the end of the 18th century was itself superseded by the English school, some of the finest mapping of the period being done by English surveyors in North America.

Herman Moll, a Dutchman, came to London about 1680 and died in St. Clement Danes in 1732, at first engraving for other publishers, then setting up on his own account, publishing separate maps and complete atlases. His works, other than his county maps, include:

A General Atlas without title, *c.* 1700.
Atlas Manuale, 1709, 8vo (43 maps).
New and Complete Atlas, 1719 (26 maps).
Atlas Minor, 1729 (62 maps).
Atlas Geographus, 5 vols., 4to, 1711–17 (100 maps and plates).

He also compiled maps for the *Complete Geographer* (1701), Oldmixon's *British Empire in America* (1708), Salmon's *Modern History* (1744–6), and Simpson's *Agreeable History* (1746). Moll's large atlas is usually bound in a tall narrow folio, the maps being folded in four. These maps measure on the average 40 by 24 inches, and are clearly printed with large cartouches enclosing the title and dedication, and many with large vignettes. Similar atlases were issued by Senex, Overton, Grierson and Willdey, the last being a toymaker as well as a map-seller.

John Senex was the contemporary of Moll, and his maps are very similar in form, but less ornate. He issued his *English Atlas* in 1714 with C. Price, later joined by John Maxwell. In 1721 he printed a *New General Atlas of the World* (34 maps). This, though smaller than the former, is still in folio size, and the titles are surrounded with vignette cartouches of more delicate appearance. Senex also issued a corrected edition of Ogilby's roads in 2 vols., 8vo, in 1719 (with 100 plates). Several editions were issued, it being continued after his death (in 1740) by his widow in 1742, 1757, 1759, 1762 and 1775. A French edition appeared in 1759 and a pirated edition by Thomas Kitchin.

A notable map was published by Henry Popple, *Map of the British Empire in America*, 1733 (20 sheets and a key map). The maps are engraved by W. H. Toms. Popple's map was easily the best that had appeared up to its date, and was reproduced abroad. It is rare, and was issued plain and coloured.

Emanuel Bowen issued a *Complete System of Geography*, 1744–7, illustrated with 70 maps, and a *Complete Atlas or Distinct View of the Known World*, 1752, with 68 maps, 20 of which relate to America. He also supplied the maps for Harris's *Voyages*, 1744, and for the *Universal History of the World*. He had the unusual distinction of being appointed engraver of maps to both George II and Louis XV. Bowen's maps have usually small but quite pleasing vignette title-pieces.

A notable exponent of the art of map-making was John Rocque, of Huguenot extraction. His best-know ork is his large-scale plan of London on twenty-four sheets in 1746, and his *Environs of London* on sixteen sheets in 1746, reissued 1748 and 1751. He likewise engraved large-scale city and county maps of Bristol (1750), Shropshire (1752), Middlesex (1754), Dublin (1756), Dublin (County) (1760), Armagh (1760), Berkshire (1761), Surrey (1765) (issued posthumously), and a charming little miniature atlas *Collection of Plans of the Principal Cities of Great Britain and Ireland*, 12mo, with 22 maps and 18 plans. The rarest of the Rocque publications is, however, that issued by his widow, Mary Ann Rocque (Topographer to His Royal Highness the Duke of Gloucester) in 1765, *A Set of Plans and Forts in America*, a small 4to volume containing 30 plans engraved by P. and J. Andrews. I have only seen three copies in forty years.

One of the most prolific map publishers towards the end of the century was Thomas Kitchin. Besides many separate maps, magazine illustrations and county maps, Kitchin issued his *General Atlas* in 1773. This was re-issued in 1777, 1780, 1782, (1786), 1787, 1795, 1796, 1798, 1808 and 1810, at first by John Sayer, then Sayer and Bennett, and finally by Laurie and Whittle.

Another noted English cartographer was James Rennell, the first to map India correctly (see Chapter XI). *Cook's Voyages*, 7 vols., 4to, and atlas, was published in 1773–84, and a further immense portion of the northern and southern Pacific as well as the coasts of Australia and New Zealand were accurately mapped and the rough outline of the world finally made known. Cook's work immediately received recognition, and was copied by cartographers of all nations.

Another important map-maker was Thomas Jefferys, who, besides large-scale maps of London and various counties, compiled his *American Atlas* issued posthumously by Sayer and Bennett in 1776 (22 maps on 29 sheets); another edition with an extra map (Lake Champlain), also dated 1776, re-issued 1778; North American Pilot, 1775 (22 maps on 36 plates), and West Indian Atlas, 1775 (39 maps), 1783, and about 1794 by Laurie and Whittle (maps increased to 61). He also issued a *Description of the Spanish Islands and Settlements on the Coast of the West Indies* in 4to in 1762, with 32 maps and plans, and *Natural and Civil History of the French Dominions in North and South America*, folio, 1760, with folding maps and plans.

Jefferys died in 1771, and was succeeded in his business by William Faden, who produced some remarkably fine maps—mostly as separate publications—besides globes, city and milit y plans and atlases. Faden's catalogue, printed in 1822, lists over 350 publications. His most valuable publication is his *North American Atlas*. Usually Faden did not publish standard atlases with printed titles, but made up collections to suit his customers' requirements, and bound them without title or list of contents, though the latter is sometimes added in manuscript. The number of maps may vary from 20 to 60. The most extensive collection was probably that formerly in the possession of Henry Stevens. This had a printed title — *The General Atlas of the Four Grand Quarters of the World, London, Printed for William Faden, successor to the late Mr. Thomas Jefferys, Geographer to the King, the corner of St. Martins Lane Charing Cross 1778*. This contained no less than 154 maps on 175 sheets. I have seen another copy with the printed title containing 22 maps on 30 sheets. Faden himself advertised the atlas at a later date as follows: *A General Atlas containing 51 maps of the Four Quarters of the Globe . . . as ratified by the definitive Treaty of Paris 1815. Colombier Folio half bound in Russia, 12 guineas.* The value of such atlases varies greatly according to their

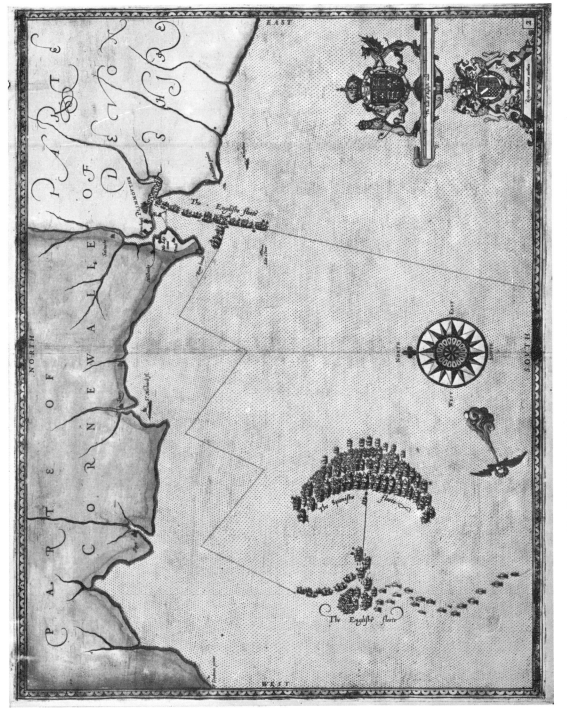

42 ROBERT ADAMS'S CHART OF SOUTHERN ENGLAND

from *Expeditiones Hispanorum in Angliam vera descriptio*, 1590. (*B.M.* 192 *f* 17 (2))

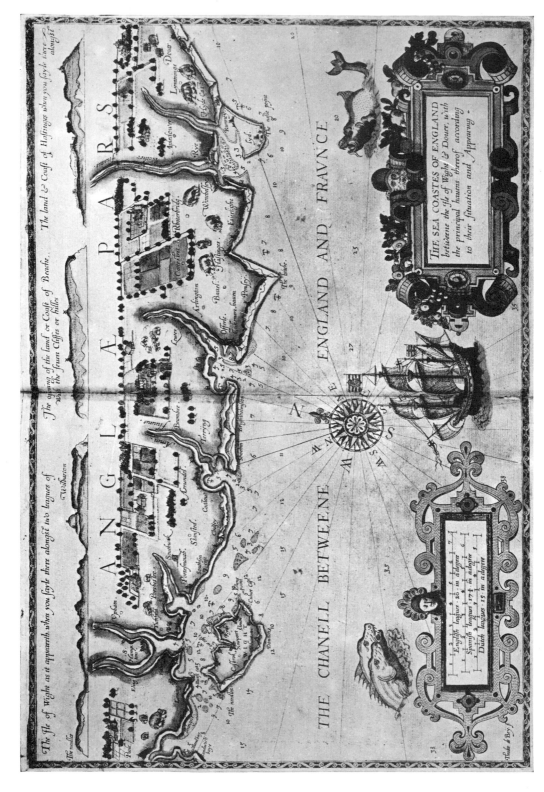

43 THEODORE DE BRY'S CHART OF SOUTHERN ENGLAND ($19\frac{1}{2}'' \times 12\frac{3}{4}''$), (DOVER TO POOLE)
from Waghenaer's *Marriner's Mirrour*, 1588. (*B.M. Maps C8 b. 4*)

contents, some of the most prized items being Ratzer's and Montresor's *Plans of New York*, and the plans of Boston, Newport and Washington's Operations.

In the 19th century map-makers became too numerous to be listed, but mention must be made of the fine large-scale maps of various parts of the world by Aaron Arrowsmith:

United States of North America, 1796, 1802, 1815 and 1819.
Chart of the West Indies and Spanish Dominions, 1803 and 1810.
New Discoveries in the Interior Parts of North America, 1795, 1796, 1802, 1811 and 1814.
South America, 1806, 1810, 1814 and 1817.
America, 1804 (4 sheets).
Egypt, 1802 and 1807.
South Africa, 1805.
Persia, 1813.
Europe, 1798.
Buenos Aires, 1806.
Mexico and adjacent provinces, 1810.
Pacific Ocean, 1798 (9 sheets).
Panama Harbour, 1800.
World, 1790 (on Mercator's projection).
World on Globular Projection, 1794 and 1827.
Asia, 1801.
Africa, 1802.
Pyrenees, 1809.

John Arrowsmith, the nephew, issued a general atlas of normal size, which went into several editions. General atlases of high quality were also issued by John Cary (1808 and 1811), Teesdale (1846), and others by Thomson (1817), Pinkerton (1815) and Lizars (1850). All these later atlases rely for their appeal (apart from their geographical interest) on the excellence and clarity of their engraving. They have no adventitious ornament, but about 1850 Tallis's *Illustrated Atlas* had small vignettes of the inhabitants, fauna and views of the country delineated, delicately engraved on each map (88). This was one of the last of the decorative atlases, though Fullarton's *General Atlas* had some maps with lithograph vignettes.

John Cary was one of the most prolific, and by many considered to be the finest, of English map-makers. His work was of high standard, and in the many issues of his works he made constant additions and improvements, not only in his geographical data, but also in technical processes of printing. Besides his maps of the British Isles, his plans of London, large- and small-scale county maps, he issued specialised maps on canals, post-roads, and separate maps of the various countries and parts of the world. The firm of G. and J. Cary also produced Guide and Road Books, Globes, Celestial Charts, Planetariums, Geological Maps, Optical and Mathematical Instruments, Microscopes, Magic Lanterns, Orreries, Air Pumps and Electrical Machines. A list of his charges is not without interest:

General Atlas of the World (61 sheets), half bound, fully coloured	£10 10s. 0d.
—— in outline colouring	£9 0s. 0d.
New Universal Atlas, Imperial Quarto, half-bound Russia	£4 14s. 6d.
—— calf	£4 10s. 0d.
—— boards	£4 4s. 0d.
Four-sheet Map of the World (5 feet 10 inches by 3 feet 2 inches)	
In sheets	£1 1s. 0d.
On rollers	£1 15s. 0d.
—— varnished	£2 4s. 0d.
—— in black and gilt frame	£4 4s. 0d.

One of the greatest figures of the 19th century was William Smith, geologist and engineer, 1769–1839. As Wright was the first to work out the mathematical formulæ for Mercator's projections, Halley to invent the meteorological and magnetic chart, so Smith was the first to produce the geological map. His great work—a Geological Map of England and Wales with Part of Scotland on 15 large sheets on a scale of 5 miles to the inch—was published August 1, 1815. This atlas is now extremely rare.

In one respect England was well served: as befitted a maritime nation, she had a long and unrivalled succession of marine atlases. At first dependent on translations from the Dutch, she quickly entered this field, mastered it, and finally excelled in it, this great effort culminating in the magnificent work of Cook, Des Barres' *Atlantic Neptune* and the Admiralty charts.

AUTHORITIES

BAKER (J. N. L.). "The Earliest Maps of H. Moll" (*Imago Mundi II*, 1937).

BRITISH RECORDS ASSOCIATION. *Proceedings No. 4*, London, 1939. (Contains article by Lynam on *Development of symbols, lettering, ornament and colour on English maps*; and *Catalogue of Loan Exhibition of Maps and Plans at Mercers Hall*, 14th November, 1938.)

CARY (G. & J.). *Catalogue of Maps, Atlases, Globes and other Works published by G. & J. Cary*, London, 1800.

CHAPMAN (S.). "Edmond Halley as Physical Geographer and the Story of his Charts." Reprinted from *Occasional Notes*, No. 9, 1941.

COLONIAL OFFICE. *Catalogue of Maps, Plans and Charts*, 1910.

CRASTER (Sir E.). "Elizabethan Globes at Oxford" (*Geogr. Jnl.*, Vol. CXVII, 1951).

CRONE (G. R.). "John Green. Notes on a 18th Century Geographer and Cartographer" (*Imago Mundi VI*, 1949).

DARLINGTON (I.), HOWGEGO (J.). *Printed Maps of London* (London, G. Phillip and Son, 1964).

Dictionary of National Biography, 22 vols., 1908–9.

EYLES (V. A.). "On the different issues of the first Geological map of England and Wales" (*Annals of Science*, 1938).

FLOWER (R.). "Laurence Nowell and the Discovery of England in Tudor Times" (*Proc. Brit. Acad.*, Vol. XXI, 1935).

FORDHAM (Sir H. G.). *Notes on the Cartography of the Counties of England and Wales* (Hertford, 1908).

—— *John Cary, Engraver and Map Seller* (Cambridge, 1910).

—— "Descriptive Catalogue of Maps" (*Trans. Biblio. Soc.*, Vol. XI, London, 1912).

—— *Notes on British and Irish Itineraries and Road Books* (Hertford, 1912).

—— *Studies in Carto-Bibliography* (Oxford, 1914).

—— An address on *The Evolution of the Maps of the British Isles* delivered in Whitworth Hall, Univ. of Manchester, 1923.

—— *Notes on the Hundred and Manor, or Grange, of Odsey* (Hertford, 1923).

—— *Road Books and Itineraries of Great Britain, 1570–1850* (Cambridge, 1924).

—— "Paterson's Roads, his Maps and Itineraries, 1738–1825" (*Trans. Bib. Soc. Library, N.S.*, Vol. V, 1925). Reprinted Oxford Univ. Press, 1925.

—— "John Ogilby, 1600–1676, His Britannia and the British Itineraries of the 18th Cent." (*Trans. Bib. Soc. Library, N.S.*, Vol. VI, 1925). Reprinted Oxford Univ. Press, 1925.

—— *Roads on English and French Maps at end of 17th century* (Southampton, 1926).

—— "Saxton's General Map of England and Wales" (*Geogr. Jnl.*, Vol. LXVII, 1926).

—— *Maps: their History, Characteristics and Uses* (Cambridge, 1927).

—— "The Earliest Tables of the Highways of England and Wales, 1541–1561" (*Trans. Bib. Soc. Library, N.S.*, Vol. VIII, 1927).

—— "A Note on the Quartermasters' Map, 1644" (*Geogr. Jnl.*, Vol. LXX, 1927).

—— "Some Surveys and Maps of the Elizabethan Period remaining in MS.; Saxton, Symonson and Norden" (*Geogr. Jnl.*, Vol. LXXII, 1928).

—— "Christopher Saxton of Dunningley, his Life and Work" (*Thoresby Soc. Miscellanea*, Vol. XXVIII, 1928).

—— *Some Notable Surveyors and Map Makers of the 16th, 17th and 18th Centuries and their Work* (Cambridge, 1929).

FORDHAM (Sir H. G.). *The Work of John Cary and his successors.* Introductory Note to the catalogue of the exhibition of Atlases, Maps, Itineraries and other Geogr. Pub. of John Cary . . . now on view at the House of the R.G.S.

GOBLET (Y. M.). "Les Cartes anglaises manuscrites des XVI et XVII siècles de la Bib. Nationale de Paris" (*Bull. Sect. de Géogr.*, Paris, 1934).

GOUGH (R.). *British Topography*, 2 vols., 1780.

HEAWOOD (E.). "Early Maps of Great Britain . . . Aegidius Tschudi's Maps" (*Geogr. Jnl.*, Vol. LXXXI, 1933).

HODGES (C. W.). "John Speed's Theatre" (*Theatre Notebook*, Vol. 3, 1949).

JOHNSTON (W. and A. K.). *One Hundred Years of Map Making* (1924).

KENNY (C. E.). *The Quadrant and the Quill, a book written in honour of Capt. Samuel Sturmy*, 1947.

LEADER (J. T.). *Life of Sir Robert Dudley*, Florence, 1895.

LYNAM (E. W. O'F.). "Woutneel's map of British Isles" (*Geogr. Jnl.*, 1933).

—— "Early Maps of the Fen District" (*Geogr. Jnl.*, 1934).

—— *Map of British Isles of 1546*, Jenkintown, 1934.

—— "Map of England by Petrus Plancius" (*B.M. Quarterly*, 1935).

—— "Robert Adams' Map" (*Geogr. Jnl.*, 1936).

—— *Atlas of England and Wales* (Saxton). Intro. by Lynam, 1936. Revised 1939.

—— *Middle Level of Fens and its Reclamation . . . with Maps of Fenland*, 1936.

—— "Early English Road Books" (*B.M. Quarterly*, 1937).

—— "The Development of Symbols, Lettering, Ornament and Colour on English Maps" (*British Records Assoc. Proc.*, 1939).

—— "Flemish Map Engravers in England in 16th Century" (*Marine*, 1943).

—— *British Maps and Map-Makers*, 1944.

—— "The character of England in Maps" (*Geogr. Mag.*, 1945).

—— "Richard Hakluyt and his successors" (*Hakluyt Soc.*, 1946).

—— "Ornament, Writing and Symbols on Maps, 1250–1800" (*Geogr. Mag.*, 1946).

—— *William Hack and the South Sea Buccaneers*. Orion Booksellers, 1948.

—— "English Maps and Map-makers of the 16th Century" (*Geogr. Jnl.*, Vol. CXVI, 1950).

MACLEOD (M. N.). "Evolution of British Cartography" (*Endeavour Apl.*, 1944).

MANLEY (G.). "Observations on early cartography of the English Hills" (*Cong. Int. Geogr.*, Amsterdam, 1938).

MAUNDER (A. S. D.). "The Origin of the Symbols of the Planets" (*Observatory*, Vol. LVIII, 1934).

MILLS (Col. D.). *Univ. of Manchester. Catalogue of Historical Maps*, arranged by Col. Dudley L. Mills, 1937.

NORTH (F. J.). *Geological Maps, their History and Development, with special reference to Wales*, Cardiff, 1928.

—— *Maps, their History and Uses, with special reference to Wales*, Cardiff, 1933.

—— "Geology's debt to Henry Thomas de la Beche" (*Endeavour*, Nov. 1944).

ORDNANCE SURVEY. Palmer (Capt. H. S.). *The Ordnance Survey of the Kingdom* (London, 1873).

—— Close (Col. Sir Charles). "The Early Years of the Ordnance Survey" (*Inst. Royal Engineers*, Chatham, 1926).

PARSONS (E. J. S.). "Wright Mollineux map of world" (*Bodleian Q. Record*, 1937).

—— and W. F. MORRIS. "Edward Wright and his work" (*Imago Mundi III*, 1939).

PELHAM (R. A.). "Early Maps of Great Britain, the Gough Maps" (*Geogr. Jnl.*, Vol. LXXXI, 1933).

PERRIN (W. G.). "The Prime Meridian" (*Mariners' Mirror*, 1927).

PHILIP (I. G.). "Early Maps of the British Isles" (*Antiquity*, 1937).

REEVES (E. A.). "Halley's Magnetic Variation Charts" (*Geogr. Jnl.*, Vol. LI, 1918).

REINHARD (W.). *Zur Entwicklung des Kartenbildes der Britischen Inseln bis auf Mercators Karte vom Jahre 1564*, Zichopau, 1909.

ROBINSON (A. H. W.). *Marine Cartography in Britain* (Leicester, 1962).

SAXTON. "The Later Editions on Saxton's Maps," by H. Whitaker (*Imago Mundi III*, 1939).

SCIENCE MUSEUM (S. Ken.). *Exhibition of Charts and Maps illust. the cartography of the British Empire*, 1928.

Sheppard (T.). *William Smith, his Maps and Memoirs*, Hull, 1920.
—— *The Evolution of Topographical and Geological Maps*. Reprinted British Association, Cardiff, 1920.
Shirley (R. W.). *Early Printed Maps of the British Isles*. Map Collectors' Circle, 1973–74.
Skelton (R. A.). "Bibliographical Notes. Pieter van den Keere" (*Library*, Sept., 1950).
Smith (W.). *Particular Description of England, 1588*, ed. Wheatley and Ashbee, 4to, 1879.
Sprent (F. P.). "The Beginnings of the Road Map in Europe" (*Roadmaker*, Vol. V, 1926).
Taylor (E. G. R.). "A Regional Map of the Early XVI Cent." (*Geogr. Jnl.*, 1928).
—— *Tudor Geography, 1485–1583*, London, 1930.
—— *Late Tudor and Early Stuart Geography, 1583–1650*, London, 1934.
—— "Robert Hooke and the cartographical projects of late 17th Century" (*Geogr. Jnl.*, 1937).
—— "Early Maps and Tide Tables" (*Antiquity*, 1938).
—— "Early attempts at precise mapping of England and Wales under auspices of Royal Soc."
     (*Con. Int. Geogr.*, Amsterdam, 1938).
—— "Hudson's Strait and the Oblique Meridian" (*Imago Mundi III*, 1939).
—— "The English Atlas of Moses Pitt *1680–3*" (*Geogr. Jnl.*, 1940).
—— "Notes on John Adams and contemporary map makers" (*Geogr. Jnl.*, 1941).
Varley (J.). "John Rocque" (*Imago Mundi V*, 1948).
[Vaughan (Rev. T.)]. *The Italian Biography of Sir Robert Dudley*, 1849.
Victoria and Albert Museum. *Tapestry Maps, 16th and 17th Century* (1915).

ENGLISH MARINE ATLASES

1588   Waghenaer (L. J.). *The Mariner's Mirrour*, folio (45 charts).   (43)
1588   Adams (Robert). *Charts of Spanish Armada* (42).
1612   Blaeu (W. Jansz). *Light of Navigation.*
          Other editions 1620 and 1622.
1625   Blaeu (W. Jansz). *Sea Mirror.*
          Another edition 1635.
1643   Blaeu (W. Jansz). *Sea Beacon.*
          Another edition 1653.
1646–7 Dudley (Sir R.). *Dell' Arcano del Mare*, Firenze, folio. The first marine atlas in which
          all the charts were drawn on Mercator's projection.
1667   Goos (Pieter). *The Lighting Colomne or Sea Mirrour* (60 maps on 59 sheets).
1667   Goos (Pieter). *Sea Atlas of the Watter-World*, folio (40 charts).
1668   Goos (Pieter). *The Sea Atlas or the Watter-World*, Amsterdam (41 charts).
          Another edition 1670, with 40 charts.
1668   Jacobsz (T.). *The Lighting Colomne or Sea Mirrour*. Amsterdam, J. & C. Lootsman,
          1668 (60 charts), 1676, and 1689.

English Pilot, 1671–1803, by John Seller.

This work had a long life, new editions constantly appearing as the work was revised and
    brought up to date; sometimes the plates were altered, sometimes entirely new plates engraved.
    The work started with Seller, then passed to Thornton and Mount, then Mount and Page,
    and the late issues to Mount and Davidson.
No library has anything like a complete set, and most of the volumes are rare, particularly the
    early issues, and the following list most probably is not exhaustive.
The editions are as follows:
English Pilot, Part I, Northern Navigation (1671) ⎫
—— Part II, Southern Navigation (1672) ⎬ 66 charts.
                                                    ⎭
A third part was issued in 1675, but not completed at the time, finishing on page 24 (7 charts).
Book III, *Oriental Navigation*, 24 pp., 17 maps, 1675. Lonsdale copy (F. E.).
The British Museum also possesses the Fourth Book, 2nd Part, dated 1671.
For the following editions the numeration of Parts I and II was changed, Part I forming the
    Southern and Part II the Northern Navigation:

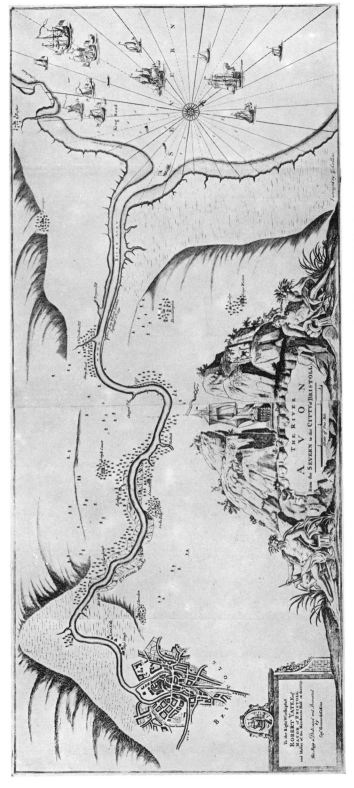

44 CAPTAIN GREENVILE COLLINS'S CHART OF THE AVON (36" × 15½")

from Great Britain's Coasting Pilot, 1693

45 DEDICATION FROM ONE OF CAPT. GREENVILE COLLINS'S CHARTS IN HIS
Great Britain's Coasting Pilot 1693

46 TITLE-PIECE FROM EMAN. BOWEN'S *SUSSEX* IN HIS
Large English Atlas c. 1760

Part I, *Southern Navigation*, by John Seller, 1690 (29 charts, B.M.).
Other editions: 1701 (27 charts, H.S.), 1702 (31 charts, B.M.), 1704 (30 charts, F.E.), 1715 (31 charts, L.C.), 1718 (24 charts, L.C.), 1738 (22 charts, H.S.), 1743 (22 charts, B.M.), 1744 (22 charts, H.S.), 1745 (22 charts, H.S.), 1747 (23 charts, L.C.), 1751 (22 charts, H.S.), 1752 (22 charts, M.), 1758 (18 charts, L.C.: 22 in H.S. and F.E.), (1760) (20 charts, B.M.), 1764 (21 charts, L.C.), 1776 (22 charts, H.S.), 1779 (22 charts, H.S.), 1785 (22 charts, H.S.), 1790 (18 charts, L.C.), 1792 (20 charts, L.C.). Also Dublin, 1772 (27 charts, H.S.).

Part II, *Northern Navigation*, 1716 (34 charts, L.C.; also B.M.).
Other editions: 1708 (Admiralty Lib.), 1723 (31 charts, L.C.), 1740 (31 charts, H.S.), 1743 (31 charts, H.S.), 1746 (33 charts, F.E.), 1749 (31 charts, H.S.), 1752, 1756 (31 charts, H.S.; also B.M.), 1770 (B.M.), 1775 (31 charts, L.C.).

Part III, *Mediterranean Sea*, by John Seller, London, John Darby, 1677 (19 charts, H.S.).
Other editions: 1703 (18 charts, B.M.), 1718 (22 charts, B.M.; pub. Amsterdam, Keulen), 1729 (15 charts, B.M.), 1733 (15 charts, H.S.), 1736 (16 charts, L.C.; 17, H.S.), 1750 (17 charts, M.), 1753, (1760) (18 charts, B.M.) 1786 (17 charts, L.C.), and 1803 (pub. by Smith and Venner, successors to Mount and Davidson, 17 charts, B.M.).
The B.M. also possesses the 2nd Book, 2nd Part, of 1677, published by J. Seller, W. Fisher, J. Thornton and J. Atkinson.

Third Book. *Oriental Navigation*, by John Thornton. Printed J. How, 1703 (35 charts, H.S.; also B.M.).
Other editions: 1711 (33 charts, B.M.), 1716 (39 charts, L.C.), 1723 (39 charts, F.E.), 1734 (41 charts, H.S.; also B.M.), 1743, 1748 (41 charts, L.C.), 1750 (42 charts, H.S.; also B.M.), 1761 (42 charts, H.S.; also B.M.).

Fourth Book. *West India Navigation*, by W. Fisher and J. Thornton, 1689; (19 charts, B.M.; 22 charts, F.E.).
Other editions: 1698 (John Thornton and Richard Mount, 19 charts, H.S.), 1706 (24 charts, L.C.), 1725 (24 charts), 1728 (21 charts, B.M.), 1737 (25 charts, L.C.; 22, H.S.; also B.M.), 1745 (29 on 34 sheets, L.C.), 1748 (22 charts, H.S.), 1749 (25 charts, L.C.; also B.M.), 1751, 1753 (25 charts, B.M.), 1755 (25 charts, L.C.; also B.M.), 1758 (25 charts, L.C.; also B.M.), 1760 (25 charts, L.C.; also B.M.), 1764 (23 charts, H.S.), 1765 (22 charts, H.S.), 1767 (22 charts, L.C.), 1770 (22 charts, H.S.), 1773 (22 charts, L.C.), 1775 (20 charts, L.C.), 1778 (21 charts, H.S.), 1780 (21 charts, L.C.), 1784 (24 charts, L.C.), 1789 (22 charts, H.S.), and Dublin, 1749 (19 charts, H.S.), 1767 (22 charts, L.C.; also B.M.), 1768

Part V, *West Coast of Africa from Straits of Gibraltar to Cape of Good Hope*, by J. Seller and C. Price. London, J. Mathews, J. Seller and C. Price, 1701 (21 charts, B.M.).
Other editions: 1720 (20 charts, H.S.), 1739 (19 charts, H.S.; also B.M.), 1744 (15 charts, M.), 1757 (19 charts, F.E.), (1760) (20 charts, B.M.), 1761, 1766 (19 charts, L.C.), 1780 (19 charts, F.E.).

Abbreviations: B.M.=British Museum; F.E.=Francis Edwards; L.C.=Library of Congress; M.=Maggs; H.S.=Henry Stevens.

1672   Seller (J.). *Coasting Pilot*, 23 charts.
1675   Seller (J.). *Description of the Sands, Shoals, Buoyes, Beacons, Roads, Channels and Sea Marks on the coast of England, South Foreland to Flamborough Head.* Folio.
1675   Seller (J.). *Atlas Maritimus.*
There are several copies of Seller's *Atlas Maritimus*, which should not be confused with Seller's *English Pilot* of 1671–2, and no two copies agree as to contents, the atlases apparently being made up to suit individual requirements (88).
The dated issue of 1675 was printed by J. Darby for the author. The L.C. copy has 42 maps and 7 coloured plates. I have seen other copies with 20, 30 and 43 charts respectively.
Of the undated issues the L.C. has 4 copies, with 26, 19, 28 and 51 charts, the first two have the imprint "Sold by James Atkinson." Others noted with 31 and 32 charts and one published *c.* 1695 Sold by Wm. Fisher and R. Mount, 19 charts.

1682   Seller (John). *Atlas Maritimus*. Sm. 8vo. (31 charts).

1683   Robijn (J.). *Zee Atlas* (English text, Dutch titles, 20 charts). Published in Amsterdam for English market.

(1685) Thornton (J.). *Atlas Maritimus or the Sea Atlas*. Folio.

1689   Robijn (J.). *The New Enlarged Lightning Sea Colom* (137 charts). Re-issued 1725.

(1690) Seller (J.). *Hydrographia Universalis or a Book of Maritime Charts*, Sm. 4to (52 charts).

1692   Jacobsz (T.). *Lightning Colom of the Midland Sea*. Amsterdam, C. Lootsman (21 charts). (19 charts in L.C. copy).

1693   Collins (Capt. Greenvile). *Great Britain's Coasting Pilot*. Folio. Sold by Richard Mount (47 charts and silhouettes) (46).
        Other editions: 1701, 1723, 1738, 1744, 1749, 1753, 1756, 1760, 1761, 1763, 1767, 1774, 1781, 1785, 1792.

(1700) Lea (P.). *Hydrographia Universalis or the Sea Coasts of the known parts of the World*. Sm. oblong folio (100 charts and 5 plates).

(1700) Thornton (J.). *Atlas Maritimus or the Sea Atlas* (23 charts in L.C.).

1703   Adair (J.). *Description of Sea Coast of Scotland* (6 charts). Part I (all published).

1708   *Atlas Maritimus Novus, or the New Sea Atlas*. Folio. Mount and Page (22 charts).

(1710) Cassini (M.). *Hydrographia Gallia: the Sea Coasts of France*. 8vo. Morden and Lea (43 charts).

1723   Norris (J.). *Compleat Sett of New Charts, North Sea, Baltick*. Folio. T. Page and W. & F. Mount (20 charts, L.C.). Another edition, 1756.

1728   *Atlas Maritimus et Commercialis*. Folio, 52 charts (and 2 star charts). Published anon. Probably work of J. Harris, J. Senex and H. Wilson.

1737   *Coasting Pilot*. Mount and Page (15 charts).

1748   Morris (L.). *Plans of Harbours*. Revised edition, 1801.

1750   *Atlas Maritimus Novus or the New Sea Atlas*. Mount and Page (20 charts). Another edition, 1755 (20 charts).

1750   Mackenzie (M.). *Orcades, or a Geographic and Hydrographic Survey of the Orkney and Lewis Islands*. Folio (8 charts). Other editions, 1767, 1776, 1791.

1753   Chart of N. and S. America, including the Atlantic and Pacific Oceans. T. Jefferys (6 sheets).
        With 4to text remarks in support of the New Chart by J. Green.

1758   Herbert (W.). *New Directory for the East Indies*. Folio (30 charts). Third edition, 1767 (37 charts).

1761   Jefferys (T.). Description of the Maritime Parts of France. Oblong folio (87 charts and plans).

1766   Speer (Capt. J. S.). *West India Pilot* (13 charts). Second edition, 1771 (charts increased to 26).

(1767) Alagna (J. G.). *Complete Set of New Charts . . . of Portugal and the Mediterranean*. Folio. Mount and Page (38 charts).

1770   Cook (J.) and M. Lane. *Collection of Charts of Coast of Newfoundland and Labrador*. London, T. Jefferys.

1772   Brahm (W. G. de). *Atlantic Pilot*. 8vo, 3 maps.

1773–88 Cook (Capt. James). (*Three Voyages*) 8 vols., 4to, and folio atlas of charts and plates.

1775   Jefferys (T.). *North American Pilot* (22 charts on 36 plates).

1775–6 *North American Pilot for Newfoundland, Labradore, the Gulf and River of St. Lawrence*. Folio, Sayer and Bennett, 1775 (22 charts). Part the Second . . . *for New England, New York, Pensilvania, Maryland and Virginia, also the two Carolinas and Florida*. Folio, Sayer and Bennett, 1776 (10 charts).

1775   *North American Pilot*.
        Another edition, 1783–4, Sayer and Bennett. 2 vols. (35 charts).
        New edition, Part I. Laurie and Whittle, 1799.
        —— Part II —— 1795 (17 charts).
        —— Part II —— 1807.

1775–81 *East India Pilot or Oriental Navigator*, 2 vols., folio. Sayer and Bennett (108 charts).

1777–86   *Complete Channel Pilot.* Sayer and Bennett (20 charts). 1779 (21 charts), 1781 (28 charts).

1777   *Atlantic Neptune.*

Published for the Royal Navy at the expense of the British Government, by J. F. W. Des Barres. There were numerous issues from 1777 to 1781, and few copies agree as to contents. The work consists, not only of charts, but also includes a varying number of views often finely tinted like water-colour drawings; though these mainly occur in the later issues, they form the most valuable collections. The plates were altered from issue to issue, some were discarded and others introduced.

The whole work forms one of the finest collections of charts and plans and views ever published. Printing started in 1774, the *Atlantic Neptune* appearing in 1777.

The largest number of copies is contained in the Library of Congress, which possesses no less than 16 examples, with the following number of charts and views: (1) 157, (2) 101, (3) 101, (4) 93, (5) 96, (6) 95, (7) (Vol. II only) 21, (8) 105, (9) 63, (10) 74, (11) 74, (12) 74, (13) (Vols. I and II only) 56, (14) (Vols. II and III only) 40, (15) 74, (16) (Vol. III only) 34.

The copy with the largest number of individual items (a collected edition) is that in the Greenwich Museum, which has 780 items (i.e. 182 separate charts and views extended to 780 items by the inclusion of different issues of the various plates). Another example formed by Henry Stevens contained 179 separate charts and views extended to 491 items by inclusion of variant plates, and a third copy with 171 charts. Another copy, Francis Edwards, 140 charts.

1778   *Neptune Occidental. A Pilot for the West Indies.* Sayer and Bennett (13 charts on 16 plates). B.M.

1779   *A New and Accurate Chart of the West India Islands and Coast.* Folio (chart and 40 plans).

(1781)   *East India Pilot or Oriental Navigator.* Sayer and Bennett (108 maps, plans and charts).

1782   Jefferys (T.). *Neptune Occidental. A complete Pilot for the West Indies.* Folio. Sayer and Bennett (25 charts on 28 plates).

1784   *Oriental Pilot.* Sayer and Bennett. Folio.

1786   Stephenson (J.) and G. Burn. *The Channel Pilot*, 2 charts, tide table. Folio. 1791 and 1803.

1787   Dunn (S.). *A New Directory for the East Indies.* Sixth edition. Folio.

1788   Robertson (G.). *Charts of the China Navigation.* Folio, 6 charts.

(1792)   Jefferys (T.). *A Complete Pilot for the West Indies.* R. Sayer.

1793   *Le Petit Neptune Français or French Coasting Pilot.* 4to. London, W. Faden (42 charts). Other editions, 1805 and 1811 (Dessiou).

(1793)   *Leard's Pilot for Jamaica and the Windward Passage.* S. Smith, successor to Mount and Davidson (15 charts).

1794–6   Bishop (Capt.). *Charts of the Gulf and Windward Passage, Old Straits of Bahamas,* Folio, (6 charts).

1795   Knight (Capt. J.). *General Chart of the Mediterranean.* Faden.

1795   —— *Coast of Portugal and Spain.* Faden.

1795   —— *Coast of Spain, France and Italy.* 6 sheets.

1795–1801   Heather (William). *A Pilot for the Atlantic Ocean.* Folio (9 charts).

1798   Arrowsmith (A.). *Chart for the Pacific Ocean.* Folio (9 charts).

1798   *Complete East India Pilot or Oriental Navigator.* Laurie and Whittle, 2 vols. (115 charts.) Other editions: 1800 (120 charts) and 1806.

1798   *Coasting Pilot for Great Britain and Ireland.*

1800   *The Oriental Pilot or East India Directory.* Laurie and Whittle (48 charts).

(1800)   *The Greenland Pilot, being Three Charts for the Fisheries of Greenland and Davis Straits.* Folio. London, David Steel.

1801   *Bougard. The Little Sea Torch . . . with corrections and additions by J. T. Serres.* 20 coloured plates of coastal views and 24 coloured plans on 12 plates. Folio.

1801   Heather (W.). *New Set of Charts for Harbours in the British Channel* (5 sheets).

1802   Knight (Capt. J.). *Charts of the Bay of Brest* (4 sheets).

1802   Heather (W.). *New Mediterranean Pilot.* 4to (224 plans, L.C.).

1803   *The Complete East India Pilot or Oriental Navigator.* Folio. Laurie and Whittle. 2 vols.

(147 charts).

1806 Arrowsmith (A.). *Pilot from England to Canton.* Folio (7 charts on 23 sheets).

1808 Heather (W.). *Marine Atlas or Seaman's Complete Pilot* (50 charts, including flags).

1811 —— *Coasting Pilot for Western Seas of Great Britain.*
    The New North Sea Pilot. 8vo.
    Sailing Directions for the River St. Lawrence.
    The New North American Pilot.
    New West India Pilot.
    New Pilot for Brazils.

1811 Hurd (Capt. T., R.N.). *Chart of English Channel.* Folio. 2 vols. (48 charts).

1814 Flinders (M.). *Voyage to Terra Australis.* 4to. 2 vols and atlas of 39 maps and plates.

1823 Smyth (Capt. W. H.). *Hydrography of Sicily, Malta and the adjacent Islands.* Folio (6 charts).

1830 Norie (J. W.). *Complete British and Irish Coasting Pilot.* Folio (16 charts).

1833 —— *Country Trade or Free Mariners Pilot.* Folio (26 charts).

1845 —— *Complete British and Irish Coasting Pilot* (23 charts).

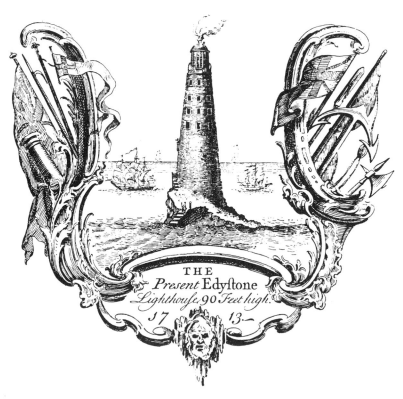

THE *Present* Edyſtone *Lighthouſe* 90 *Feet high.* 1713.

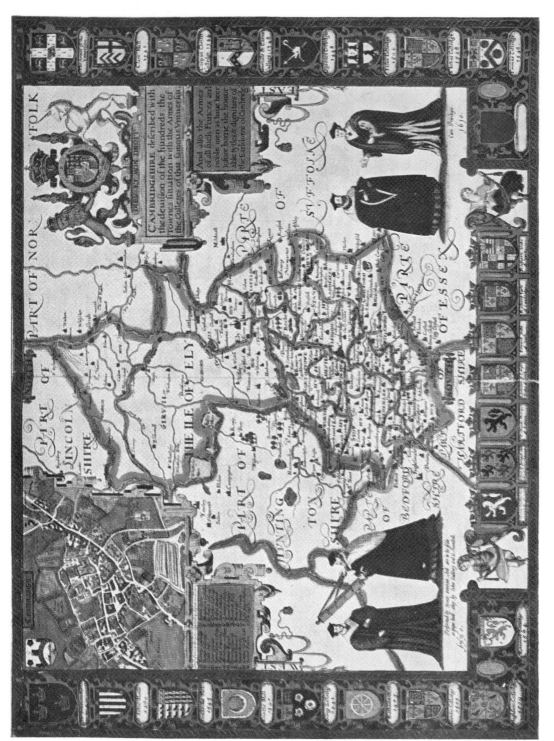

47   JOHN SPEED'S CAMBRIDGE (20″ × 15″)
from *The Theatre of the Empire of Great Britain*, 1611

CHAPTER VIII

The County Maps of England and Wales

THE county maps of England and Wales have an ancient history commencing in the 16th century, the first English atlas in this field appearing before its counterpart in France, and in succeeding centuries a constant and increasing stream of publications was issued on our national topography. They present a wide and varied choice to the collector.

ELIZABETHAN ENGLAND (*16th Century*)

The oldest known series of regional maps is that compiled by Laurence Nowell (1559–1576). They remained unprinted; the manuscript is in the British Museum. Another manuscript in the possession of Lord Lansdowne contains a much finer example of Nowell's cartographic artistry.

Saxton's Maps, 1579

The first printed and most important map of any English or Welsh county is that of Christopher Saxton. Saxton, a Yorkshireman, was born about 1542. He became attached to Thomas Seckford, Master of the Court of Requests and Surveyor of the Court of Wards and Liveries. At this gentleman's expense, and under the authority of Queen Elizabeth, he surveyed and drew maps of all the counties of England and Wales, in which task he was engaged for many years. Saxton's work is all the more extraordinary when compared with the work of later cartographers. In surveying the whole of England and Wales in detail, he exceeded in extent the work of any other cartographer of his own, or in fact, other lands. His maps are excellent in every way—valuable as being the first printed map of any county; important as original work; accurate considering the means at his disposal, and well produced. They set a standard and remained the base for succeeding county maps for over a hundred years. They were finely engraved by Ryther, R. Hogenberg, N. Reynolds, Terwoort and Scatter, and are strikingly decorative. They were issued both plain and hand-coloured, and in contradistinction to the usual practice, the uncoloured copies are more rare than the coloured. Contemporary coloured examples with their emblazoned coats-of-arms, their balanced interwoven and festooned strapwork, ships and small figures, are amongst the most pleasing of English county maps (38). Original impressions of Saxton's maps are rare, and cost from £100 upwards according to the county desired, but excellent coloured reproductions were printed by the British Museum at 12s. 6d. each. Unfortunately, most—it is hoped only temporarily—are out of print. Saxton's maps are variously dated from 1574 to 1579, his complete atlas being published at the latter date. As printed maps of Elizabethan England they are highly prized, and Saxton has rightly been called the "Father of English Cartography." Saxton's maps and a few by Norden were reproduced a few years later on a reduced scale in the 1607 edition of Camden's *Britannia*, and this was re-issued in 1610 and 1637. Later still they were reprinted in their full size by William Webb in 1645, and again by Philip Lea in 1689 and 1693, by Willdey in 1720, and as late as 1749 by Thomas Jefferys. Some of these later editions bore alterations.

The most notable of Saxton's later works was his magnificent large-scale map of England and Wales. Engraved on 21 sheets, it was on a scale of about 8 miles to the inch. Its existence was known for a long time owing to the revision published by Lea in 1687 (which incidentally was altered), but no copy of the original edition was known right up to 1930, when Messrs. Francis Edwards found a complete copy. This passed into the hands of the British Museum. In 1938 the same firm discovered a second example of this precious map, differing from the former in that it was enclosed within a broad engraved border containing 83 coats-of-arms of the nobility and gentry. This now reposes in Birmingham Public Library. This map was used on a reduced scale, and also in detail, for most of the succeeding maps of England till well into the 18th century.

Philip Lea secured the copper-plates, printed a fresh title and considerably altered the plates, changing the coats-of-arms of the nobility and wording and adding the principal roads which he took from Ogilby's survey of 1675. He also changed the Elizabethan ships to a Carolinian aspect. This reprint is likewise rare.

John Norden, 1548–1625

John Norden, a contemporary of Saxton, is the next greatest figure of Elizabethan cartography. He was born in 1548 and died about 1625. His birth-place is unknown, but according to his own statement it was in Somerset. It was evidently his intention to make an independent survey of the counties, but, less fortunate than Saxton through lack of financial assistance, he failed. Even the work he did accomplish was only printed in part in his lifetime (so far as is known), though his material was used by other cartographers. The following of his cartographical works are known:

1593   Descriptive account of Middlesex with map of county. This was published under the title *Speculum Britanniae*, Part I.

1594   Map of Surrey, 16 by 13 inches, engraved by Charles Whitwell. Two copies known, in B.M. and R.G.S. Library. A re-issue of about ten years later is known from a solitary copy in the British Museum.

1595   Map of Sussex, 20½ by 11¼ inches, engraved by Christopher Schwytzer. Only known copy in R.G.S. Library.

c. 1595   Map of Hamshire.

1598   Account of Hertfordshire, with map of the county.

Norden likewise made other surveys that remained in MS. during his lifetime: Cornwall (not published till 1720); Essex (first published by Camden Society in 1840); and Northants. He is also credited with being the author of the map of Kent in the 1607 edition of Camden's *Britannia*, which suggests the possibility of an earlier original that has yet to be found. Since a distinguishing feature of Norden's maps is the insertion of main roads (he was apparently the first to do this, none being shown on Saxton), he may also have made a map of Warwickshire, as this map in the Anonymous series of 1602–3 shows the roads. Owing to the rarity of Norden's maps, he is best known perhaps for his *Guyde for English Travellers*, 1625, comprising triangular distance tables, a system invented by Norden. He is credited with the following maps in the Camden *Britannia* of 1607: Hampshire, Surrey, Sussex, Kent, Hertfordshire and Middlesex. The excellence of Norden's work may be gauged from the fact that his contemporaries seem to have preferred his maps where available.

48   BLAEU'S MAP OF BUCKINGHAM (10½″ × 16½″)
from Part IV of the *Theatrum Orbis Terrarum*, 1645

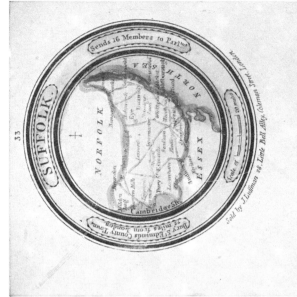

SUFFOLK is 62 miles in length from east to west, and 28 in breadth from north to south. It is divided into 22 hundreds, which contain 28 market towns, 575 parishes, 32,805 houses; and the population amounts to 210,431.

The air is clear and healthy, even near the sea-coast; the soil of various qualities, and the country in general level. Its products are grain of all sorts, peas, beans, turnips, carrots, and hemp; butter and cheese, and excellent draught horses. On the whole, this county, in respect to agriculture, is one of the most thriving in the kingdom. The principal rivers are the Stour, the Waveny, the Little Ouse, the Deben, and the Orwel.

50 J. LUFFMAN'S SUFFOLK (DIAMETER 2¼")
from *A New Pocket Atlas*, 1803

BARK-SHIRE.

HUNDREDS In *Bark-shire*.

1. Orner.
2. Ganfield.
3. Farington.
4. Shrivenham.
5. Wanting.
6. Compton.
7. Morton.
8. Lamborne.
9. Faircrofse.
10. Theale.
11. Reading.
12. Charleton.
13. Sonnynge.
14. Wargrove.
15. Barnerthe.
16. Braye.
17. Ripplemore.
18. Cookham.
19. Oke.
20. Kentbury.

A

ABINGTON, *Hor.*
Aldermerfton, *The.*
Aldworth, *Compt.*
Apleford, *Oke.*
Apleton, *Oke.*
Arberfield, *Sen.*
Ardington, *Want.*
Afhamfteed, *Mort.*
Afhbury, *Shriu.*
Afton upthorp, *Mort.*
Afton

49 T. JENNER'S BERKSHIRE (4" × 4"). [The 1657 issue]
from *A Direction for the English Traveller*, 1635.

Minor Map-makers

A few other minor names of this period may be lightly mentioned—minor, not on account of their workmanship, but from the smallness of their output in this field. Philip Symonson composed a map of Kent. This was printed on two large sheets and engraved by Charles Whitwell. No copy of the first issue is at present known, but a sheet of the eastern half is in the possession of Canon Livett, and this bears the date 1596. Heawood calls it "Perhaps the finest specimen of English cartography before 1600," and it vies with Norden's efforts in showing the main roads. Symonson was a remarkably fine draughtsman, and several manuscript plans of his are in existence, notably some of the Sussex coast. Another fine manuscript map of Kent is that by Robert Glover, Somerset Herald, in 1571.

Another worker in the field of cartography was William Smith, Rouge Dragon. Many of his MSS. and other heraldic works are in existence. His cartographical works are as follows:

1585   Map of Cheshire, with a descriptive account, which includes a folding view of Chester and a small view of Haulton. The map is a corrected version of Saxton, and was used in the Anonymous Series and also by Speed.
1588   Description of England illustrated by a general map and plans of various cities. A reproduction of this MS. was printed in 1879, edited by H. B. Wheatley.
1598.   Map of Lancashire, differing in detail from Saxton, on which it is based.

Elizabethan Plan-makers

Less known and far more rare than the county maps, a large number of surveys were made of the principal towns of the kingdom prior to 1600. In fact, some antedate the county maps. Some exist only in manuscript, like John Walker's fine plan of Chelmsford, 1591; William Smith's bird's-eye views of Bristol, 1568; Salisbury and Canterbury, 1588; Norwich and London; and in the manuscript plans in Lord Burleigh's copy of Saxton's atlas, Falmouth Haven, Shrewsbury and Scarborough Castle; Robert Norman's Mouth of the Thames, 1580; William Norton's Platte of Tottenham Courte, 1591; and John Blagrave's Forest and Manor of Feckenham, 1591. Others were printed as William Cunningham's plan of Norwich, 1559; Richard Lyne's plan of Cambridge, engraved 1574; Hamond's plan of Cambridge, 1592; Agas's Oxford, 4 by 3 feet, 1578–88; Hooker's Plan of Exeter, 20 by 13¾ inches, engraved by R. Hogenberg 1587 (of which there are three issues, only one example of each being known). In 1573 Braun and F. Hogenberg commenced their great work, the *Civitates Orbis Terrarum* in five volumes, a sixth volume being added much later in 1618. This work included plans of London, Cambridge, Oxford, Norwich, Bristol, Chester, Edinburgh, Canterbury, Exeter, York and Dublin.

Many remarkably fine estate maps date from this period, notably some by Saxton, Norden, Walker, Symonson, Thomas Langdon and William Senior.

Anonymous Map-maker of 1602–3

A new series of county maps was projected in 1602–3, but as far as is known the series was never completed. They are with one exception without editor, publisher or engraver's name. Based in the main on Saxton, both Norden and Smith appear to have been utilised in their compilation: Norden in Surrey, Herts, Northants and possibly Essex, and William Smith in Lancashire and Cheshire. They are in the Dutch Flemish style with little ornamentation, save for somewhat simple title-pieces. They are extremely rare, and even the late reprints of them by Overton and Stent are uncommon. Twelve of these maps only are known:

Essex, dated 1602, bearing the name Hans Woutneel. Most like Saxton's, but with roads added, probably after Norden, and with minor alterations.

Leicester and Rutland, dated 1602. Based on Saxton, but with no less than 80 additional names supplied by William Burton, the county historian.

Warwick, dated 1603. On a larger scale than Saxton, with the addition of 60 more place-names. Roads are shown.

Surrey, copied from Norden.

Hertfordshire, copied from Norden, but on a larger scale.

Northants, credited to Saxton, but most probably Norden.

Staffordshire, north to right of map.

Suffolk.

Norfolk, Cheshire, Lancashire, Worcestershire.

The Royal Geographical Society possesses originals of the first seven on the above list, and the British Museum has an original of Suffolk. The last four are known so far only from later re-issues by Peter Stent, published between 1650–70.

Camden's Britannia, 1607

In 1607 an edition of Camden's *Britannia* was published. This was the last Latin edition published in Camden's lifetime, and the first to have a series of county maps. The text was in Latin. The work was re-issued in 1610, and 1637 with the same maps, but the text translated into English. These maps, engraved by William Kip and William Hole, are in the main reductions of Saxton's maps of 1579, but where available the maps of Norden were employed, the following being credited to him: Hampshire, Surrey, Sussex, Kent, Hertfordshire and Middlesex. The map of Pembrokeshire is by George Owen. They measure on the average 14 by 11 inches (engraved surface). The decorative title-pieces, scales of miles and other ornaments were newly designed, and these maps, by their convenient size, their charming appearance and their antiquity, make most pleasing additions to any county collection. It is worth noting that, apart from the set of playing cards of 1599, the 1607 edition of Camden is remarkable for giving the first separate printed representation of the following counties: Bedfordshire, Berkshire, Oxford, Bucks, Middlesex, Cambridge, Huntingdon, Lincoln, Nottingham, Rutland, Cumberland, Westmorland, the Ridings of Yorkshire, and all the counties of Wales except Glamorgan and Pembroke. In Saxton's atlas the above counties were grouped together in two or more counties to a sheet.

England under the Stuarts (17th Century)
John Speed's Maps, 1610–11

We now come to the best known and most popular of all English county maps, namely, those by Speed. John Speed, born 1552 at Farndon in Cheshire, was a worthy successor to Saxton, industrious, painstaking and persevering. He issued his *Theatre of the Empire of Great Britain* in 1611, though some of the maps were issued before that date, and undoubtedly sold separately; many bear the date 1610. These early issues have plain backs, the whole series not being completed till 1611, when they were issued in atlas form with printed text on the reverse, giving an account of the most notable events and principal features of the county. I have seen these early issues in various trial states. The series consists of a separate map for each county in England and Wales, besides general maps of each kingdom, one for Scotland and five for Ireland. Speed's atlas enjoyed great popularity and went into many folio editions. A miniature atlas in pocket size was also issued and this went into many editions. Speed's maps were mainly based on the work of Saxton and Norden, but he also used material from the series of Anonymous maps of 1602–3, for example Leicester, and for Kent he turned to Symonson's map for information, though in this case he gave no acknowledgment. He also included a certain amount of fresh material and introduced certain new features. His atlas was the first to show the counties divided into hundreds, and the first to give small plans or views

ATLAS

MARITIME

TITLE-PAGE TO R. DE HOOGHE'S *ATLAS MARITIME*, 1693

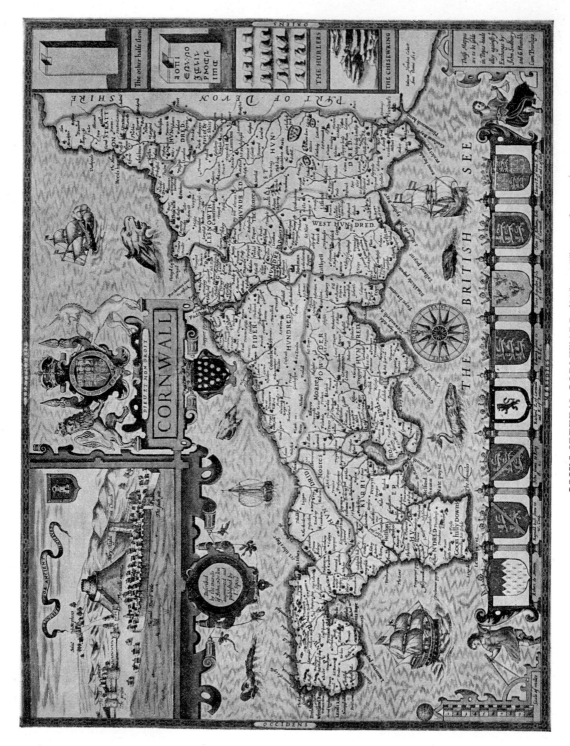

JOHN SPEED'S CORNWALL (20" × 15")

from *The Theatre of the Empire of Great Britain*, 1611

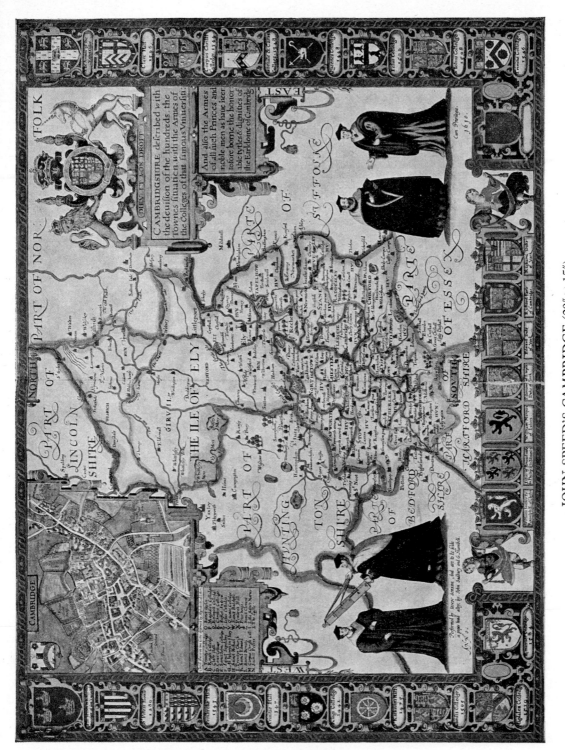

JOHN SPEED'S CAMBRIDGE (20″ × 15″)

from *The Theatre of the Empire of Great Britain*, 1611

CHRISTOPHER SAXTON'S MAP OF DORSET (21" × 15")

from his *Atlas* 1791

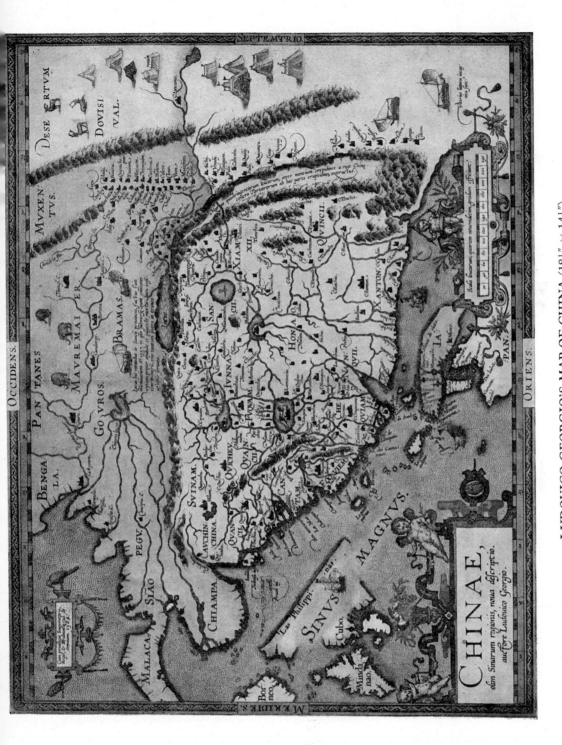

LUDOVICO GEORGIO'S MAP OF CHINA ($18\frac{1}{2}'' \times 14\frac{1}{2}''$)

from Ortelius's *Theatrum*, 1584

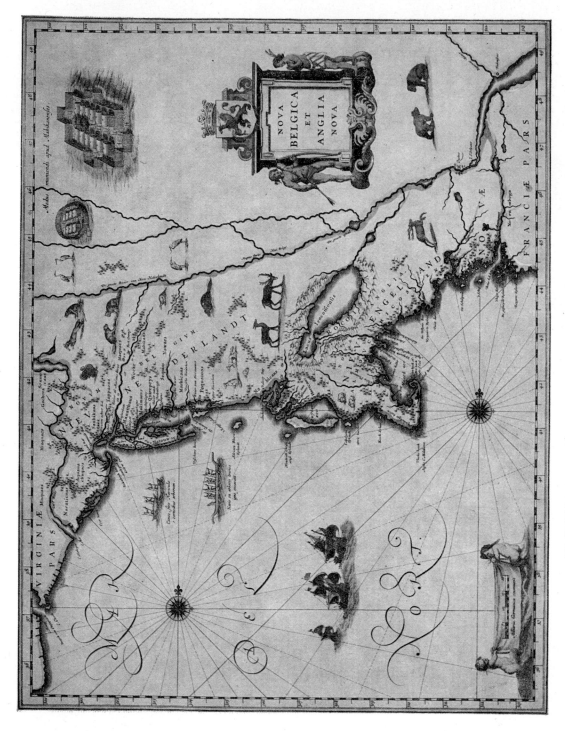

BLAEU'S NEW BELGIUM AND NEW ENGLAND (20" × 15")

from the *Atlas Major*, 1662

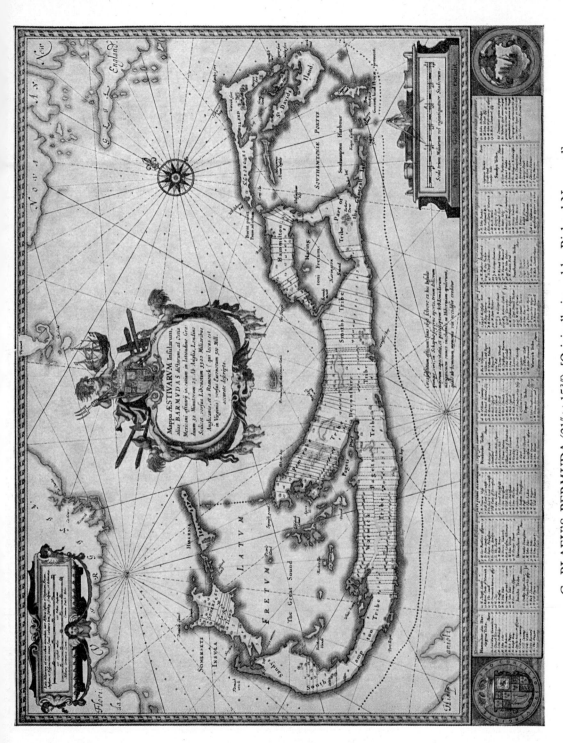

G. BLAEU'S BERMUDA (21″ × 15¾″), [Originally issued by Richard Norwood]

from the *Atlas Major*, 1662

DECORATIVE SCALE FROM JAILLOT'S MAP OF PALESTINE, 1691

of the principal towns, usually two to each map. These plans were taken from various authors —William Mathew, Christopher Schwytzer, Norden, Agas, Cunningham and Smith. The plans of Pembroke and St. David's alone were definitely claimed as by Speed himself. He likewise introduced a considerable amount of heraldry into the framework of his maps, and from the 1616 Latin edition onwards, a certain amount of classical information. He likewise had his maps superbly engraved by Jodocus Hondius the elder. They were issued uncoloured; a few contemporary owners had their examples coloured, but this was the exception rather than the rule (36, 47, 51, and 65).

Some writers have written rather disparagingly of Speed, basing their criticism on his lack of originality. This is hardly fair, and possibly not even correct. It is true that Speed is not as great a man as Saxton, and that his work is mainly derivative, but he went to the best sources for his initial framework and gave due acknowledgment where possible—which

51 *Speed's Miniature Map*, 1627

cannot be said of many later geographers—and his innovations were many. He discarded Saxton, where he found later and better information, and his maps were constantly revised in one particular, place-names. It is only in recent years that intensive research has begun on this facet of his maps, the otherwise excellent bibliography of Chubb not even having noted the fact that there are small alterations and additions in the succeeding editions of his work. Finally, it may be said that Speed introduced much that was of value and interest into his maps, stimulated an interest in his subject, and produced a work of beauty.

The early issues of Speed's maps bear the imprint of the publishers, Sudbury and Humble (Humbel or Hüble). Later issues have the names of Roger Rea, Basset and Chiswell, Overton and Dicey. Chubb lists no less than fourteen folio editions of Speed's atlas, but since his time a great deal of research has been done, notably by Mr. Whitaker and the late Dr. Gardner, and the list has been considerably extended. All the editions are worth collecting owing to their variations, some of the later ones, particularly Dicey's imprint, being rare. Speed is the first, as far as I know, to give a separate printed map of the Isle of Wight (by William White), the Isle of Man (by Thos. Durham), and the four provinces of Ireland.

Drayton's Maps, 1612–22

In 1612 Michael Drayton published his *Polyolbion or Chorographical Description of all the Tracts, Rivers, Mountains, Forests and other Parts of the Renowned Isle of Great Britain.* A second part was issued in 1622. They are the queerest and most fantastic series of English county maps. No county boundaries are given and hardly any details. As implied in the title, rivers are shown; and out of these rivers spring water nymphs and titular deities. On the hills and Downs large figures of shepherds and shepherdesses are placed; forests are shown with Noah's Ark trees, and countrymen disport in the plains. Neptune rides in the sea, and symbolical figures crowned with castle and spires represent the occasional towns. The maps are not of great geographic value, but have a pleasing decorative appeal, with their blending of the formal and the free, their mixture of stiff Stuart costume and pagan nudity. They are engraved by William Hole (40).

Blaeu's Maps, 1645

In 1645 J. Blaeu published his first atlas of the counties of England and Wales. It forms Part IV of his *Theatrum Orbis Terrarum*. Perhaps to some it may not make so great an appeal as the work of the older English cartographers, Saxton and Speed—the slightly archaic character is gone. Blaeu still shows monsters in his sea; but one has the feeling that they are used purely for decoration rather than as a matter of belief. On the other hand, there are compensations: everything is as calculated and exact as possible. Typographically they are superb, the paper is the finest procurable, the calligraphy magnificent, the impressions always clean and sharp and beauty is not forgotten. The cartouches surrounding the individual titles have become larger, and usually include some figures giving an indication as to the pursuits or industry of the county depicted; the scales of miles are treated decoratively and the arms of the principal nobility depicted, maritime counties having the addition of sailing ships and sea monsters. No shoddy work was ever allowed to proceed out of the establishment of the Blaeus'. For these maps Blaeu took his information mainly from Speed and Saxton, but their presentation bears the stamp of his own genius (48). There were several issues with varying text, French, Dutch, German, Italian and Spanish, as well as English.

Jan Jansson, 1646

Jan Jansson, or to give him his Latin name Joannes Janssonius, was the great rival of Blaeu. Jansson published his atlas of England and Wales in 1646. In style and decoration they are similar to those of Blaeu, in some cases being identical except for the imprint. The title to each map is enclosed within a graceful framework of symbolical figures, indicating the typical pursuits and attributes of the county. They are further adorned with the royal arms and heraldic shields, showing the arms of the nobility and county families. The counties bordering the sea are further adorned with ships and compass roses. Jansson's maps, like Blaeu's, were issued coloured and uncoloured. The contemporary coloured copies are the most desirable. This first issue has Latin text on the back of each map. Re-issued 1646 and 1647 (French text), 1647 and 1649 (Spanish text), 1652 (no text), 1652 (Dutch text), 1659 and 1666 (Latin text). Further editions of Jansson's maps were published by Schenk and Valck, and Allard, up to 1724.

RESTORATION ENGLAND

Blome's Maps, 1673

Richard Blome issued a series of county maps in his *Britannia*, 1673. From a geographical point of view they have little merit, being rather sketchy affairs, but from the decorative point

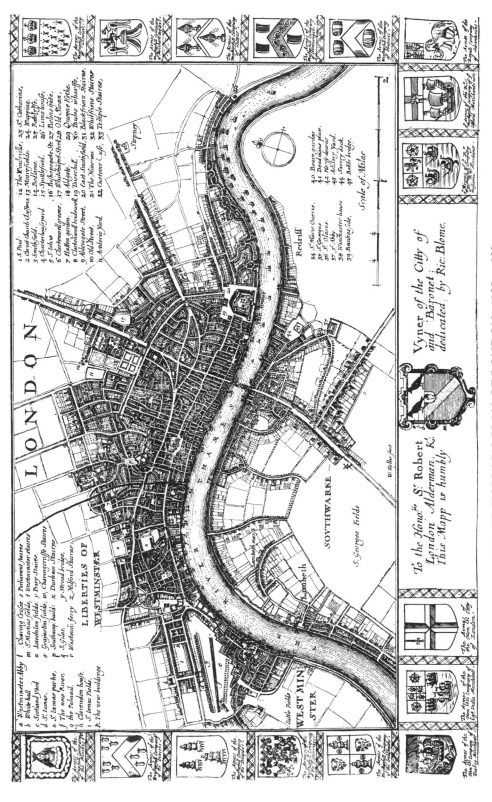

52 W. HOLLAR'S PLAN OF LONDON (10¾" × 6¼")

from Blome's *Britannia*, 1673

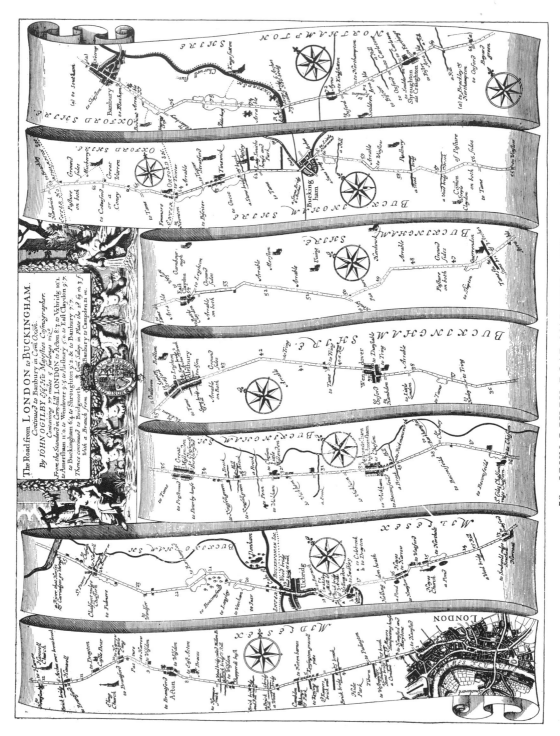

53 JOHN OGILBY: ROAD MAP (17½" × 13")
from his *Britannia*, 1675 the earliest road maps of England

of view they are not without charm, with their quaint little flourishing title-pieces and ornamental dedications (52).

AUGUSTAN ENGLAND
Morden's Maps, 1695

During the reign of William and Mary, Robert Morden produced a series of county maps for Gibson's edition of Camden's *Britannia*. They were described as "all new engraved either according to surveys before publish'd or according to such as have been made and printed since Saxton and Speed." They were sent to "Knowing Gentlemen" in each county to be rectified, and are finally described as "by much the fairest and most correct of any that have yet appear'd." In spite of this contemporary puff, their general appearance is still in that somewhat coarse and free style characteristic of the early map-makers. They are greatly inferior to the maps of Blaeu both in execution and decoration. They may be called decorative, as the title of each county is within an ornamental cartouche. Moreover, they are the only large county maps of this period (54). Morden was fairly prolific. He also issued a series of county maps on a smaller scale.

HANOVERIAN ENGLAND
Bickham, 1754

In 1754 curious representations of the counties of England were issued by George Bickham. They are hardly maps, but rather bird's-eye perspective views, the county being shown in relief, with attractive figures in the foreground depicting the costume of the period (59).

Bowen and Kitchin, 1750–80

An important series of county maps was published by Bowen and Kitchen. They are the best and most attractive county maps of the 18th century. Emanuel Bowen was appointed engraver of maps to George II and Louis XV. The most notable feature of his maps is the great amount of space devoted to descriptive text in the form of historical notes spread over their surface. They are peculiar in this respect, and in spite of these notes are clear, well engraved and not unattractive. Several of the counties have quite pleasing vignette views of the principal towns, and the title-piece of each county is always within an elaborately designed cartouche, with rococo ornament, costume figures, heraldic coats of arms, etc. (58).

Cary and Smith

At the end of the 18th and beginning of the 19th century, John Cary issued several atlases of the counties of England. They are entirely bare of ornament, but impressive in their workmanship. They are remarkable for their accuracy and their clear, clean print; indeed, they are brilliantly engraved, and rank with the Ordnance survey as the finest maps of the nineteenth century. From a visual, as distinct from a decorative point of view, they have not been surpassed. Cary has been described by Sir George Fordham as "the most representative, able and prolific of English cartographers." This is high praise and not unmerited, but Cary was not the only excellent English cartographer of the period, even in the sphere of English County Maps (60), for Charles Smith produced equally able work in his *New English Atlas* of 1804, and this also went into many editions.

Greenwood, 1834

Charles and James Greenwood were the most ambitious of the later map-makers. They published a beautifully engraved *Atlas of the Counties of England* mostly from their own surveys. Each map is embellished with a large vignette of a notable building in the county. Some were issued uncoloured, but the finest examples were printed in colour, and make extremely handsome maps. Their more important work was the issue of a series of large-scale county maps. A. Bryant did similar fine work.

VICTORIAN ENGLAND

Moule's Maps, 1836

Strictly speaking, Thomas Moule's maps were first published in the reign of William IV, but they continued to be issued during the reign of Victoria, later editions having her portrait as a frontispiece. They are clearly printed and attractive, engraved on steel, the last series of decorative maps to be published. Each county is enclosed within an ornamental border, with symbolical or costume figures woven into the design, heraldic coats of arms, and from two to three or four vignette views of notable scenes in the county (63).

For further information on the maps of the counties of England see Chaps. VIII*a* and *b*.

AUTHORITIES

BEDFORD. *Catalogue of the Enclosure Awards, Supplementary Catalogue of Maps and List of Awards upon Tithe, in Bedfordshire County Muniments*, Bedford, 1939.

BERKSHIRE. *National Register of Archives in conjunction with Berks. County Council, Cat. Exhib. Art Gallery, Town Hall, Reading*, 1951.

BUCKINGHAMSHIRE. Price (Ursula). "The Maps of Buckinghamshire, 1574–1800" (*Records of Buckinghamshire*, Vol. XV, Pts. 2 and 3, 1948–9).

—— Schulz (H.). *An Elizabethan Map of Wotton Underwood*. San Marino, Cal., 1939.

CAMBRIDGESHIRE. "John Hamond's Plan of Cambridge, 1592," by John Willis Clark (*Proc. of Cambridge Antiq. Soc.*, Vol. VII, 1888).

—— Fordham (Sir H. G.). "Cambridgeshire Maps, 1579–1800" (*Proc. of Cambridge Antiq. Soc.*, 1905–7–8).

—— *Cambridgeshire Maps: a Descriptive Catalogue, 1579–1900* (Cambridge, 1908) (limited to 100 copies).

—— Clark (J. W.) and A. Gray. *Old Plans of Cambridge, 1574–1798*, 2 parts (Cambridge, 1921).

CHARTERED SURVEYORS' INSTITUTE. Loan Exhibition of Old Maps and Plans, 1932.

CHESHIRE. Harrison (W.). "Early Maps of Cheshire" (*Trans. of Lancs. and Cheshire Antiq. Soc.*, Vol. XXVI, Manchester, 1909).

—— Whitaker (H.). "Descriptive List of the printed maps of Cheshire, 1577–1900" (*Chetham Soc.*, 1942).

CHUBB (T.). *The Printed Maps in the Atlases of Great Britain and Ireland: a Bibliography, 1579–1870*. London, 1927. Reprint, 1968.

CUMBERLAND AND WESTMORLAND. "The Chorography, or a Descriptive Catalogue of the Printed Maps of," by J. F. Curwen (*Trans. of the Cumberland and Westmorland Antiq. and Arch. Soc.*, *N.S.*, XVIII, 1918).

DEVON. "The Early Printed Plans of Exeter, 1587–1724," by K. M. Constable. (*Reprinted from Trans. of Devonshire Association*, 1932.)

ESSEX. "Description of the County of Essex by John Norden, 1594." Edited from original MS. by Sir H. Ellis (*Camden Society*, 1840).

—— "John Norden's Map of Essex," by William Cole (*Essex Naturalist*, Vol. I, 1887).

—— Avery (J.). "Christopher Saxton, Draughtsman of the oldest-known map of Essex" (*Essex Naturalist*, Vol. XI, 1901).

—— Huck (T. W.). "Some Early Essex Maps and their Makers" (*Essex Review*, Vol. XVIII, Colchester, 1909).

—— Green (J. J.). "Chapman and André's Map of Essex (1777)" (*Essex Review*, Vol. XIX, Colchester, 1910).

—— *Catalogue of Maps in the Essex Record Office, 1566–1855*. Edited by F. G. Emmison (Essex County Council, 1947).

—— *The Art of the Map-maker in Essex, 1566–1860* (Essex County Council, 1947).

FORDHAM (Sir H. G.). *Notes on the Cartography of the Counties of England and Wales* (Hertford, 1908).

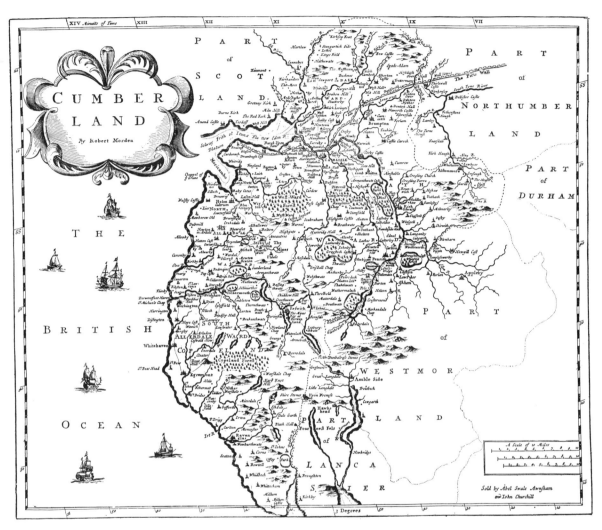

54   ROBERT MORDEN'S *CUMBERLAND* ($16\frac{1}{2}'' \times 14\frac{1}{4}''$)
from William Camden's *Britannia*, 1695

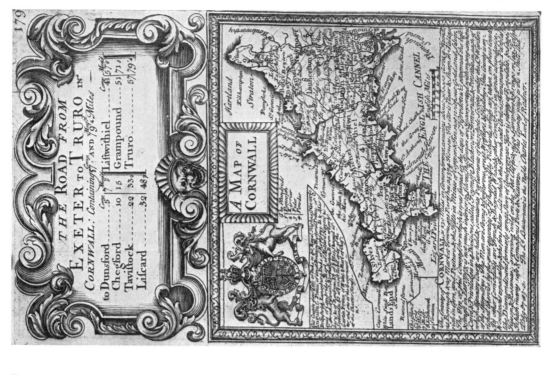

56 E. BOWEN'S *CORNWALL* ($4\frac{1}{2}'' \times 4\frac{1}{2}''$)
from *Britannia Depicta*, 1720

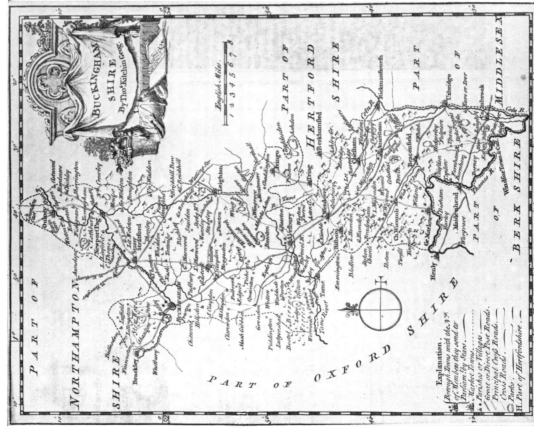

55 T. KITCHIN'S *BUCKINGHAMSHIRE* ($6\frac{1}{2}'' \times 8''$)
from *London Magazine*, 1748

GLOUCESTERSHIRE. *Some Account of the Oldest Plans of Bristol*, by William George (Bristol, 1881).
—— Pritchard (John E.). *Exhibition of Old Bristol Plans, etc., July 17th, 1906. Catalogue.*
—— Pritchard (John E.). "Plan of Bristol by Jacobus Millerd, c. 1670" (*Bristol Arch. Notes for 1909*, Bristol, 1910).
—— Chubb (T.). "Descriptive Catalogue of the Printed Maps of Gloucestershire, 1577–1911" (*Trans. of Bristol and Glos. Arch. Soc.*, 1912, Vol. XXXV, Bristol, 1913).
—— *Additions to, and Notes on the above*, by R. Austin (Bristol, 1917).
—— Pritchard (John E.). "A Hitherto Unknown Original Print of the Great Plan of Bristol," by Jacobus Millerd, 1673 (*Trans. of Bristol and Glos. Arch. Soc.*, Vol. XLIV, 1922).
—— Pritchard (John E.). "Old Plans and Views of Bristol" (*Trans. Bristol and Glos. Arch. Soc.*, Vol. XLVIII, 1927).
—— Grove (L. R. A.). "Map of the Manor of Tyersall, 1720" (*Bradford Antiquary N.S. XXXIV*, 1947).
HAMPSHIRE. Box (E. G.). "Norden's Map of Hampshire, 1595" (*Proc. Hampshire Field Club and Arch. Soc.*, 1939).
—— —— "Hampshire in Early Maps and Early Road Books" (*Proc. Hampshire Field Club and Arch. Soc.*, Vol. XIII, Pt. I).
HEAWOOD (E.). "Some Early County Maps" (*Geogr. Jnl.*, Vols. LXVIII and LXIX, 1926–7).
—— *English County Maps in the collection of the Royal Geographical Society*, 1932.
HERTFORDSHIRE. Gerish (W. B.). *The Hertfordshire Historians, John Norden, 1548–1626 (?)—a Biography* (Ware, 1903).
—— Fordham (Sir H. G.). "Hertfordshire Maps; a Descriptive Catalogue, 1579–1900" (*Trans. Hertfordshire Nat. Hist. Soc.*, 1901-3-5 and 1907).
—— —— Re-issued in 1 volume with addition of Preface, Notes and Index (Hertford, 1907) (limited to 50 copies).
HUMPHREYS (A. L.). *Old Decorative Maps and Charts*, 1926.
HUNTINGDONSHIRE. *Catalogue of Books and Maps relating to Huntingdonshire* in County Library, Huntingdon, 1950.
KENT. Hannen (Hon. H. A.). "Account of a Map of Kent dated 1596" (*Arch. Cantiana*, Vol. XXX, London, 1913).
—— —— "Further Notes on Phil Symonson" (*Arch. Cantiana*, Vol. XXXI, London, 1915).
—— Box (E. G.). "Lambarde's 'Carde of this Shyre' " (*Arch. Cantiana*, Vol. XXXVIII, 1926).
—— —— "Lambarde's 'Carde of this Shyre'." Third issue with Roads added (*Arch. Cantiana*, Vol. XXXIX, 1928).
—— —— "Notes on some West Kent Roads on Early Maps and Road Books" (*Arch. Cantiana*, Vol. XLIII).
—— —— "Two 16th-century Maps of Kent" (*Arch. Cantiana*, Vol. XLV, 1934).
—— Livett (G. M.). "Early Kent Maps (Sixteenth Century)" (*Arch. Cantiana*, Vol. XLIX, 1937).
—— Heawood (E.). "The earliest known maps of Kent" (*Geogr. Jnl.*, 1938).
—— Box (E. G.). "Kent in Early Road Books of the 17th Cent." (*Arch. Cantiana*, Vol. XLIV).
—— Campbell (E. M. T.). "An English Philosophico-Chorographical Chart [Packe's East Kent]" (*Imago Mundi VI*, 1949).
LANCASHIRE. Roeder (C.). "Maps and Views of Manchester" (*Trans. of Lancashire and Cheshire Antiq. Soc.*, Vol. XXI, Manchester, 1904).
—— Harrison (W.). "Early Maps of Lancashire and their Makers" (*Trans. of Lancs. and Cheshire Antiq. Soc.*, Vol. XXV, Manchester, 1908).
—— Whitaker (H.). "Descriptive List of the Printed Maps of Lancashire, 1577–1900" (*Chetham Soc.*, 1938).
—— Harley (J. B.). "William Yates and Peter Burdett" (*Trans. Hist. Soc. Lancs & Cheshire*, 1963).
LEICESTERSHIRE. Gimson (B. L.) and P. Russell. *Leicestershire Maps, a Brief Survey*, 1947.
LINCOLNSHIRE. Goshawk (Evelyn). "Old Lincolnshire Maps" (*Lincolnshire Historian*, No. 3, 1948).
LONDON. Crace (F.). *Catalogue of Maps, Plans and Views of London*, 1878.
—— Mitton (E. E.). *Maps of Old London*, 1908.
—— Darlington (I.) and Howgego (J.). *Printed Maps of London* (London, G. Phillip and Son, 1964).
MACLEOD (M. N.). "Evolution of British cartography" (*Endeavour*, Apl. 1944).

MANCHESTER UNIV. LOAN EXHIBITION (Col. Mills, Mrs. Booker), 1923.

NORFOLK. *Streets and Lanes of the City of Norwich.* A Memoir by John Kirkpatrick, ed. William Hudson, to which are added Early Maps of the City of Norwich, with Intro. by W. T. Bensly, Norwich, 1889.

—— "Rough Catalogue of Maps relating to Norwich and Norfolk," by Walter Rye and H. Brittain (*Norfolk Antiq. Miscellany*, Series II, Pt. I, Norwich, 1906).

—— Squire (J. C.). "Norwich in the 16th cent." (*Norwich P.L. Readers' Guide*, 1926).

—— Chubb (T.) and G. H. Stephen, *Descriptive List of the Printed Maps of Norfolk, 1574–1916: Descriptive List of Norwich Plans, 1541–1914*, Norwich, 1928.

NORTHAMPTONSHIRE. Heawood (E.). "A hitherto unrecorded MS. map of Northamptonshire by John Norden" (*Geogr. Jnl.*, 1940).

—— Whitaker (H.). "Descriptive List of the Printed Maps of Northamptonshire, 1576–1900" (*Northamptonshire Record Soc.*, 1948).

NORTHUMBERLAND. Hadcock (R. Neville). "A Map of Mediaeval Northumberland and Durham" (*Soc. of Ant. of Newcastle-upon-Tyne*, 1939).

—— Whitaker (H.). "A Descriptive List of the Maps of Northumberland, 1576–1900" (*Society of Antiquaries of Newcastle-upon-Tyne*, 1939).

NOTTINGHAMSHIRE. Wadsworth (F. Arthur). "Nottinghamshire Maps of the 16th, 17th and 18th centuries, their Makers and Engravers" (*Trans. Thoroton Soc.*, Vol. XXXIV, 1930).

OXFORD. "Old Plans of Oxford," by Agas, Hollar and Loggan (*Oxford Hist. Soc.*, Vol. XXXVIII, Oxford, 1898).

—— "Oxford Topography." An Essay by Herbert Hurst, forming a companion volume (*Oxford Hist. Soc.*, Vol. XXXIX, Oxford, 1899).

SOMERSET. Chubb (T.). *Descriptive List of Printed Maps of Somersetshire, 1575–1914* (Taunton, 1914).

STAFFORDSHIRE. (Burne (S. A. H.). "Early Staffordshire Maps" (*Trans. of N. Staffordshire Field Club*, Vol. LIV, Stoke-on-Trent, 1920).

—— "Addenda" (1926).

—— Heawood (E.). "County Map (Staffs), by William Smith, 1599" (*Geogr. Jnl.*, 1939).

SUFFOLK. Sanford (W. G.). *The Suffolk Scene in Books and Maps*, 1951.

SURREY. Sharp (H. A.). "Historical Catalogue of Surrey Maps," with a Bibliography (*Readers' Index*, Croydon P.L., Croydon, 1927–9).

—— "Rocque's Map of Surrey," by Wilfrid Hooper (*Surrey Arch. Collection*, Vol. 40, 1932).

SUSSEX. *A Survey of the Coast of Sussex made in 1587*, edited by Mark Antony Lower (Lewes, 1870).

—— Gerard (E.). "Notes on some early printed Maps of Sussex and their Makers" (*Library*, 3rd Series, Vol. VI, London, 1915).

—— —— —— "Early Sussex Maps" (*Sussex County Magazine*, 1928).

—— —— Hastings Museum, *Catalogue of Maps and Plans in the Exhibition of Local Maps*, 1936.

—— —— Steer (F. W.). "Sussex Estate and Tithe Award Maps" (*Sussex Record Soc.*, 1962).

WALES. "The Road Books of Wales with a Catalogue, 1775–1850" (*Arch. Cambrensis*, Vol. LXXXII, reprinted London, 1927).

—— North (F. J.). *Map of Wales before 1600 A.D.* (Cardiff Nat. Museum of Wales, 1935).

—— —— *Maps, their History and Uses, with Special Reference to Wales* (Cardiff, 1933).

—— Evans (O. C.). *Maps of Wales and Welsh Cartographers* (London Map Collectors' Circle, 1964).

WARWICKSHIRE. Jourdain (M.). "Notes on Tapestry Maps representing Warwickshire" (*Memorials of Old Warwickshire*, ed. A. Dryden, 1908).

WILTSHIRE. Chubb (T.). "Descriptive Catalogue of the Printed Maps of Wiltshire, 1576–1885" (*Wilts. Arch. & Nat. Hist. Mus.*, Vol. XXXVII, Devizes, 1911).

YORKSHIRE. Sheppard (T.). *The Evolution of Kingston upon Hull as shewn by its Plans* (Hull, 1911).

—— —— "East Yorkshire History in Plan and Chart" (*Trans. of E. Riding Antiq. Soc.*, Vol. XIX, Hull, 1912).

—— —— *The Lost Towns of the Yorkshire Coast* (London, 1912).

—— Whitaker (H.). "Descriptive List of the Printed Maps of Yorkshire and its Ridings, 1577–1900" (*Yorks. Arch. Soc.*, Record Series, LXXXVI, 1933).

—— Grove (L. R. A.). "Map of the Manor of Tyersall" (*Bradford Antiquary N.S.*, Pt. XXXIV, 1947).

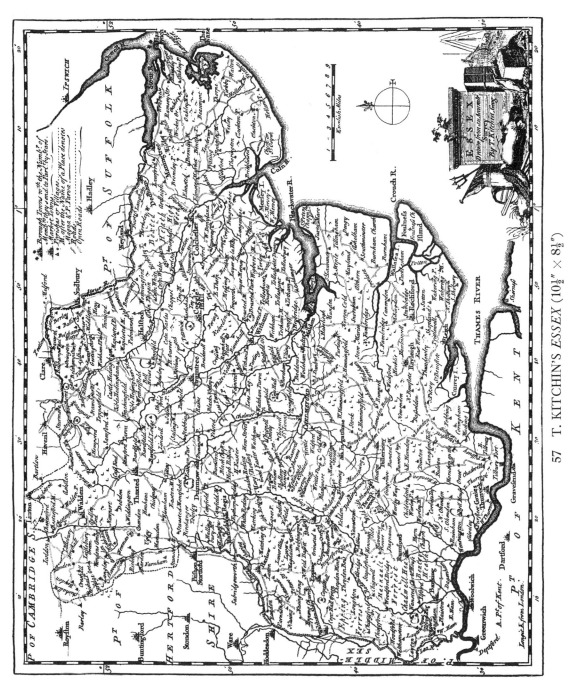

57 T. KITCHIN'S ESSEX (10½″ × 8½″)

from Boswell's *Antiquities of England and Wales*, 1786

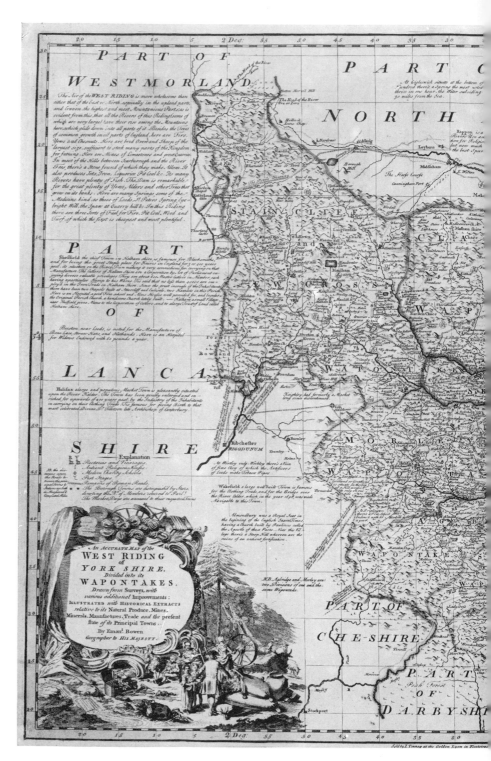

58   E. BOWEN'S WEST RIDING ($27\frac{1}{2}'' \times 20\frac{1}{2}''$)

from *The Large English Atlas*, 1749–55

59   G. BICKHAM'S BIRD'S-EYE MAP OF DEVONSHIRE (5½″ × 9″)
from *The British Monarchy*, 1750

a—LARGE-SCALE COUNTY MAPS OF ENGLAND

Bedford
*Bryant (A.). 4 sheets, 1826.
*Gordon (W.). 2 sheets, 1736.
*Greenwood (C. & J.). 4 sheets, 1826. 1 inch.
*Jefferys (T.). 8 sheets and Index, 1765. 2 inch.
—— second edition, 8 sheets, Faden, 1804.

Berkshire
*Greenwood (C. & J.). 4 sheets, 1824. 1 inch.
*Rocque (J.). 2 sheets, 1752.
*—— 18 sheets and index map, J. Rocque, 1761. 2 inch (64).
*—— —— M. A. Rocque, 1761–2.
13-page Index issued, but rarely found with map.

Buckinghamshire
*Bryant (A.). 4 sheets, 1824–5. 1½ inch.
*Jefferys (T.). 4 sheets, March 24, 1770. 1 inch.
*—— second edition, 4 sheets, 1778.

Cambridgeshire
*Baker (R. G.). 4 sheets, (1821.) 1 inch.
*—— another edition, 1830.
*Cole (B.). 520 by 620 mm., engraved J. Harris, 1710.

Channel Isles
Gardner (W.). Guernsey, 2 sheets, 1787.
—— T. Cubitt and others. Jersey, 1795.

Cheshire
*Burdett (P. P.). 4 sheets, 1777. 1 inch.
*—— second edition, 1794.
*Greenwood (C.). 4 sheets, 1819.
*Swire (W.) and W. F. Hutchings, Teesdale, 1830. 3 inches=4 miles.

Cornwall
*Gascoigne (Joel). 14 sheets, engraved I. Harris (1700).
*—— another edition, Mount and Page (1730).
*Greenwood (C.). 6 sheets, 1827.
*Martyn (T.). 7 sheets, title on 2 slips, R. Sayer, 1748.
*—— 7 sheets and title on 2 sheets, Faden, 1784.
Index printed in Bodmin in 8vo in 1816.

Cumberland
*Greenwood (C. & J.). 6 sheets, 1823.
*Donald (T.). 6 sheets, engraved J. Hodskinson, 1774. 1 inch.
*—— second edition, W. Faden, 1802.

Derbyshire
*Burdett (P. P.). 6 sheets, engraved T. Kitchin (1767). 1 inch.
*—— another edition, improved W. Snowdon, pub. J. Cary, 1791.
*Greenwood (C.). 1825. 1 inch.
*Sanderson (G.). 2 sheets, engraved J. & C. Walker, 1836.

Devon
*Donn (B.). 12 sheets and index map, engraved T. Jefferys, 1765. 1 inch.
*Greenwood (C. & J.). 9 sheets, 1827.

Dorset
  Greenwood (C. & J.). 6 sheets, 1826.
  *Taylor (I.). 6 sheets and index map, 1765. 1 inch.
  *—— second edition revised, 6 sheets, Faden, 1795–6.
  *—— Reduced 1 sheet 30 by 20 inches, 1796. ½ inch.

Durham
  *Armstrong (Capt. A.). 4 sheets, engraved Jefferys (London), J. Chapman, 1768. 1 inch.
  *—— —— R. Sayer & T. Jefferys, 1768.
  *Greenwood (C.). 4 sheets, 1820. 1 inch.
  *Hobson (W. C.). Engraved J. & C. Walker, 1840. 1⅛ mile to inch.

Essex
  *Chapman (J.) and P. André. 26 sheets, including Index, 1777. 2 inch.
  *—— another edition, Keymer, 1785.
  *Greenwood (C. & J.). 6 sheets, 1825. 1 inch.
  Oliver (J.). 1696 (Essex Record Office, Chelmsford).
  *Overton (P.) & T. Bowles. 2 sheets, 1726.
  *Warburton (J.), J. Bland and P. Smyth. Engraved S. Parker, 1749 (Middlesex, Essex and Herts).

Gloucester
  *Bryant (A.). 6 sheets, 1824.
  *Donn (B.). Eleven miles round Bristol, 4 sheets, 1769. 1½ inch.
  *Greenwood (C. & J.). 6 sheets, 1824.
  *Taylor (I.). 6 sheets, Ross, 1777.
  *—— another edition, Faden, 1786.

Hampshire
  *Greenwood (C. & S.). 6 sheets, 1826.
  *Milne (T.). 6 sheets, Faden, 1791. 1 inch.
  *Taylor (I.). 6 sheets, 1759. 1 inch.
      Isle of Wight. Andrews (J.). 4 sheets, 1769. Another edition, 1775.
          —— Albin (J.). 1795. 1 inch.
      New Forest. Richardson, King and Driver. 10 sheets and index map, Faden, 1789.

Hereford
  *Bryant (A.). 1835. 1½ inch.
  *Price (H.). 2 sheets, 1817.
  *Taylor (I.). 4 sheets, 1754. 1 inch.
  *—— 4 sheets, Faden, 1786.

Hertfordshire
  *Dury (A.) and J. Andrews. 9 sheets and index map, Faden (1766), and separate plans of Hertford
      and St. Albans.
  *—— another edition, W. Faden, 1782.
  *Oliver (J.). 2 sheets, 1695.
  Seller (J.). 20 by 16 inches, 1733.

Huntingdon
  *Greenwood (C. & J.). E. Ruff, 1830.
  *Jefferys (T.). 6 sheets, 1768. 2 inch.
  *Gordon (W.). 6 sheets, 1730–1.

Isle of Man
  Drinkwater (J.). 24 by 36 inches, 1826.
  *Fannin (P.). 20 by 28 inches, 1789.

76

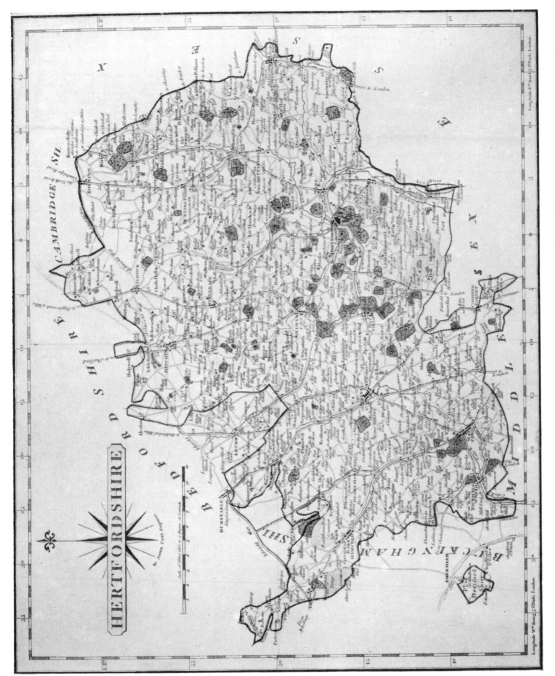

60 J. CARY'S *HERTFORDSHIRE* ($10\frac{1}{4}'' \times 8\frac{1}{4}''$)
from *New and Correct English Atlas*, 1787

61 LANGLEY'S *MAP OF KENT* (10″ × 7″)

from *New County Atlas, c.* 1818

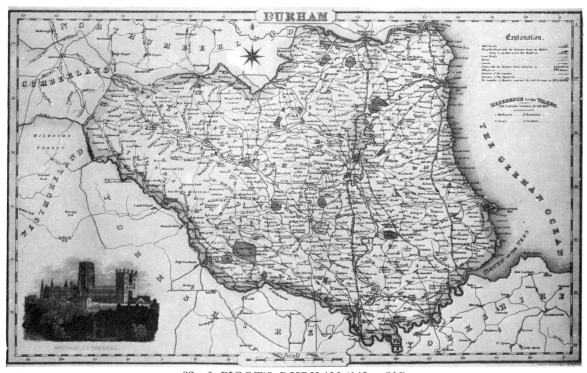

62 J. PIGOT'S *DURHAM* (14″ × 8½″)

from *The British Atlas*, 1831

Kent
    *Andrews (J.), A. Dury and W. Herbert. 27 sheets, with index map, 1769. 2 inch.
    *——— ——— ——— another edition 1775.
    *——— ——— ——— 27 sheets, Sayer and Bennett, 1779.
    ——— ——— ——— Laurie & Whittle, 1794.
    *Greenwood (C. & J.). 4 sheets, 1821.
    *Packe (C.). East Kent, 4 sheets, 1743.
    Ruff (E.). 4 sheets, 1837.
    *Seller (J.). $35\frac{1}{2}$ by 23 inches, engraved John Oliver and Richard Palmer. Sold H. Moll and P. Overton (1710).
    ——— another issue. I. King added to imprint.
    Symonson (P.). 31 by 21 inches, engraved Whitwell, 1596. Reissued Stent (1659).

Lancashire
    *Cruchley (G. F.). 6 sheets, 1836.
    *Greenwood (C.). 1818, 1 inch.
    *Hennet (G.). Teesdale, 1830. 3 inches=4 miles.
    *Yates (W.) and T. Billings. 8 sheets, 1786.
    *——— ——— second edition, 8 sheets, Faden, 1800.

Leicestershire
    *Greenwood (C. & J.). 4 sheets, 1826.
    *Prior (J.). 4 sheets, 1777.
    ——— W. Dawson, 1779. 1 inch.
    *King (W.). Country surrounding Belvoir Castle, 4 sheets, 1806.

Lincolnshire
    *Armstrong (A.). 8 sheets, 1778. An 8vo index issued, 1779.
    *Bryant (A.). 2 sheets, 1828, 1 inch.
    *Greenwood (C. & J.), 6 sheets, 1830.

Middlesex
    *Andrews (J.) and A. Dury. Sixty-five Miles round London, 20 sheets, 1777.
    *Greenwood (C.). 4 sheets, 1819.
    *Rocque (J.). 4 sheets, 1754. 2 inch.
    *Seller (J.), engraved by I. Oliver. Richard Davies, 1710.
    *——— another edition, J. Smith, 1714.
    *——— ——— 1724.
    *——— ——— 1732.
    *——— ——— Thomas Millward, 1742.
    *Warburton (J.), J. Bland and P. Smith. Middlesex, Essex and Herts, engraved S. Parker, 7 sheets, 1749.

Monmouthshire
    *Greenwood (C. & J.). 4 sheets, 1830.

Norfolk
    *Bryant (A.). 6 sheets, 1826. $2\frac{1}{2}$ inches=2 miles.
    *Corbridge (J.). 20 by 27 inches, 1765.
    *Donald (T.), Milne and others. 6 sheets, Faden, 1797.
    *Goddard and Goodman. 2 sheets, Norwich, 1731.
    *——— ——— another edition, 1740.

Northamptonshire
    *Bryant (A.). 1827.
    *Eyre (T.) and T. Jefferys. 4 sheets, 1779–80. 1 inch.
    *——— ——— second edition revised, Faden, 1791.
    *Greenwood (C. & J.). 4 sheets, 1826.

Northumberland
    *Armstrong (A.). 9 sheets, engraved Kitchin, 1769. 1 inch.
    *Fryer (J.) and Sons. 6 sheets, 1820.
    *Greenwood (C.). 6 sheets, 1828. 1 inch.
    *Warburton (J.). 4 sheets, 1716. 1 inch = 2½ miles.

Nottinghamshire
    *Chapman (J.). 4 sheets, 1776.
    *——— further editions 1785, 1792.
    *Ellis (G.). 4 sheets, engraved Foot, 1827.
    *Greenwood (C. & J.). 1826.

Oxfordshire
    *Bryant (A.). 4 sheets, 1824. 1½ inches to mile.
    *Davis of Lewknor (R.). 16 sheets and key map, 1797. 2 inch.
    *Jefferys (T.). 4 sheets, 1768-9. 1 inch.
    *——— another edition, R. Sayer (1775).
    *Overton (P.). 2 sheets, 1715.

Rutland
    *Armstrong (A.). 20 by 25 inches, 1781.
    *King (W.). Country surrounding Belvoir Castle . . . including the whole county of Rutland,
        4 sheets, 1806.

Shropshire
    *Baugh (R.). 9 sheets, 1808.
    *Greenwood (C. & J.). 6 sheets, 1827. 1 inch.
    *Rocque (J.). 4 sheets, 1752. 1 inch.

Somersetshire
    *Day and Masters. 9 sheets, 1782. 1 inch.
    *——— another edition, 1800.
    *Greenwood (C. & J.). 6 sheets, 1822.

Staffordshire
    *Greenwood (C.). 4 sheets, 1820.
    *Jefferys (T.). Smith, Newcastle and Bowles, London, 1747.
    *Phillips (J.) and W. E. Hutchings. Teesdale, 1832. 3 inches = 4 miles.
    *Yates (W.). 6 sheets, engraved J. Chapman, 1775. 1 inch.
    *——— ——— second edition, W. Faden, 1799.

Suffolk
    Bryant (A.). 1826. 1¼ inch = 1 mile.
    *Greenwood (C. & J.). 6 sheets, 1825.
    *Hodskinson (J.). 6 sheets, Faden, 1783. 1 inch.
    *Kirby (J.). 28 by 20 inches, engraved Basire, 1737.
    *——— 4 sheets, additions by J. & W. Kirby, engraved Ryland, 1766.

Surrey
    *Bryant (A.). 4 sheets, 1823. 1½ inch.
    *Greenwood (C. & J.). 4 sheets, 1823.
    *Lindley (J.) and Crossley (W.). Engraved by Baker, 2 sheets, 1790. 1 inch.
    *——— ——— another edition, 1793.
    Rocque (J.). 9 sheets, engraved P. Andrews, 1768.
    *——— ——— 1770.
    *——— third edition, engraved P. Andrews, 1775. 2 inch.
    *Senex (J.). 4 sheets, 1729. 1 inch.

*Seller (J.), J. Oliver and R. Palmer. Engraved J. Oliver and R. Palmer, 1680.
*—— another edition, P. Lea (1690).
*—— —— T. Bowles & J. Bowles, 1733.

Sussex
    *Budgen (R.). 6 sheets, engraved Senex, 1724. 3 inches=4 miles.
    *—— republished with Turnpike Roads corrected from a survey by J. Sprange . . . and R. Budgen, 1779.
    *Figg (W.). 6 sheets, 1861.
    *Greenwood (C.). 6 sheets, 1825.
    *Overton (P.). 39 by 23 inches (1740).
    *Senex (J.). Surrey, Kent, Sussex, 1746.
    *Yeakell (T.) and W. Gardner. 4 sheets, 1778–83. 2 inch.
    *——another edition . . . completed by T. Gream, engraved T. Foot, 4 sheets, Faden, 1795.

Warwickshire
    *Greenwood (C. & J.). 4 sheets, 1822.
    Yates (W.) & Sons. 4 sheets, 1793 for John Sharp. 1 inch.

Westmorland
    *Greenwood (C. & J.). 4 sheets, 1824.
    *Jefferys (T.). 4 sheets, 1770.

Wiltshire
    *Andrews (J.) and A. Dury. 18 sheets and Index, 1773.
    *—— second edition revised, 18 sheets and Index, Faden, 1810.
    *Greenwood (C.). 4 sheets, 1820.

Worcestershire
    *Greenwood (C. & J.). 4 sheets, 1822. 1 inch.
    *Taylor (I). 4 sheets, Ross, 1772. 1 inch.
    *—— second edition, 1800.

Yorkshire
    *Bryant (A.). E. Riding, 4 sheets, 1829. 1 inch.
    *Greenwood (C.). 9 sheets, 1817–18.
    *Hobson (W. C.). 2 sheets, 1843.
    *Jefferys (T.). 20 sheets and index map, 1771–2.
    *—— —— second edition, 1775.
    *—— —— third edition, corrected Rev. G. Markham and F. White, 20 sheets, 1800.
    *—— —— reduced, revised and corrected, W. Faden, 4 sheets, 1816.
    *Overton (P.). 2 sheets, 1728.
    *Teesdale (J.). 9 sheets, 1828.
    *Tuke (J.). 4 sheets 1787.
    *Warburton (J.). 4 sheets engraved S. Parker (1720).

b—ENGLISH COUNTY ATLASES

Agricultural Soc. of England (Royal) Journal 1843–70, 8vo, 36 geological maps.
Aiken (J.). *England Delineated*, 1790, 8vo, 43 maps.
    Other editions 1795, 1800, 1803, 1809.
Andrews (J.). *Historical Atlas of England*, 1797, 13 plates.
Archer (J.). *British Magazine*, 1841, 8vo, 17 ecclesiastical maps.

Badeslade (T.). *Chorographia Britannia*, 1742, 8vo, 46 maps. Engraved by W. H. Toms; maps dated 1741.

—— Two other issues in the same year, one with price added to title-page, other with imprint on maps changed to 1742.

Large paper copies exist: also 1743.

—— Second edition, 1745: another edition 1747.

Baker (Benjamin). *Universal Magazine*, 1791–8, 8vo, 44 maps.

Another edition 1807 (Laurie and Whittle's New and Improved Atlas), 47 maps.

Barclay's Dictionary, see Moule.

Bell (J.). *New and Comprehensive Gazetteer of England and Wales*, 1833–4, 4 vols., 8vo, 44 maps.

Bickham (George). *British Monarchy*, *1743–54*, folio, 48 maps (59).

—— Other editions 1749, 1796 (new title, *Collection of Bird's-eye Views of the several Counties in England and Wales*.)

Blaeu (G. & J.). (*England*) 1645, folio, 56 maps. This forms Vol. IV of the *Theatrum Orbis Terrarum*, published by Blaeu in Amsterdam (48).

Other editions 1645 (French and German text), 1646 (Latin text), 1647 (Dutch text), 1648 (five issues, Latin French, German, Dutch text and one without text), 1662 (2 issues, Latin and French text), 1663 (French text), 1664 (Dutch text), 1667 (French text), 1672 (Spanish text), 1714 (French text).

Blome (R.). *Britannia*, 1673, sm. folio, 50 maps. (There were at least three issues of this work, the dedications being changed in several of the maps) (52).

—— *Speed's Maps epitomised*, 1681, 8vo, mostly engraved by Hollar and Palmer, 39 maps: and re-issued in 1685.

—— *Cosmography*, 1693, sm. folio (with series of county maps).

Boswell (H.) and T. Kitchin. *Antiquities of England and Wales*, 1786, folio, 50 maps.

Bowen (Emanuel and Thomas). *Atlas Anglicanus*, 1767, folio, 45 maps. Another edition 1777.

Bowen (E.). *English Atlas*, 1794, folio, 43 maps.

A re-issue of Bowen's *Royal English Atlas* of 1762.

Bowen (E.) and T. Kitchin. (*Large English Atlas*) (1755), folio, published by J. Hinton and J. Tinney (58).

Other editions, *The Large English Atlas*, T. Bowles, John Bowles and Son, John Tinney and Robert Sayer, folio, 45 maps: 1763: John Bowles, Carington Bowles and Robert Sayer, folio, 47 maps (Scotland and Ireland added): 1777, Carington Bowles: 1785, Robert Wilkinson, successor to Mr. John Bowles, folio, 47 maps: 1787, Robert Sayer, folio, 50 maps.

Bowen (E.) and Kitchin, T. *The Royal English Atlas*, T. Kitchin, R. Sayer, Carington Bowles, H. Overton, H. Parker, J. Bowles and J. Ryall (1762), folio, 44 maps.

Other editions; 1778 (R. Wilkinson), 44 maps: 1780 (Sayer and Bennett), 49 maps.

Bowen (E.) and Benjamin Martin. "The Natural History of England" (*General Magazine of Arts and Sciences*), 1759–63, 8vo, 40 maps.

Bowles (Carington). *Bowles's New Medium English Atlas*, 1785, 4to, 44 maps.

—— *Bowles's Pocket Atlas*, 1785, folio, 57 maps.

Butters (R.). *An Atlas of England*, 1803, 8vo, 40 maps.

Camden (William). *Britannia*, 1607, sm. folio, 57 maps.

Maps after Saxton, Norden, Owen, and Anon. map-maker of 1602–3. Engraved Kip and Hole. Latin text on reverse of maps.

—— Re-issued 1610 (reverse of maps blank), sm. folio, 57 maps.

—— Another edition 1637 (reverse of maps blank), sm. folio, 57 maps.

—— 1695, 1715, 1722, 1755, 1772. See Morden (R.).

—— *Britannia* 1789, 3 vols., folio, 60 maps (engraved J. Cary).

—— Abridgment of Camden's *Britannia*, 1626, sm. 8vo, 52 maps.

Capper (B. Pitts). *Topographical Dictionary of the United Kingdom*, 1808, 8vo, 44 maps.

Other editions 1813, 1825, 1826, 1829.

Cary (J.). *New and Correct English Atlas*, 1787, 4to, 46 maps (60).
     Other editions 1793 (several issues), 1809, 1812, 1818, 1821, 1823, 1825, 1826, 1827, 1831 (taken over by Cruchley, 1862).
—— *Cary's English Atlas*, 1809, folio, 42 maps.
     Other editions 1811, 1818, 1828, 1834.
—— *Cary's Travellers' Companion*, 1790, 8vo, 43 (road) maps.
     Other editions 1791, 1806, 1810, 1812, 1814, 1817, 1819, 1821, 1822, 1824, 1826, 1828.
Cary (J.) and John Stockdale. *New British Atlas*, 1805, folio, 43 maps.
Cassell (Petter and Galpin). *Cassell's British Atlas*, 1864–7, folio.
Clarke (B.) and H. G. Collins. *British Gazetteer*, 1852, 8vo, 43 maps.
Cobbett (W.). *Geographical Dictionary of England and Wales*, 1832, 8vo, 52 maps.
     Another edition 1854.
Cole (G.) and J. Roper. *The British Atlas*, 1810, 4to, 58 maps and 21 plans.
Collins (H. G.). *Travelling Atlas of England and Wales*, 1850, 8vo, 45 maps.
     Other editions 1852, 1868.
Cooke (G. A.). *Modern British Traveller*, 1802–10, 25 vols., 8vo, 46 maps.
     Another edition 1822.
Cowley (J.). See Dodsley (R.).
Cruchley's County Atlas of England and Wales, 1862, 8vo, 46 maps.
     Other editions 1863, 1868, 1872, 1875.
Cruchley (G. F.). *Railway Companion to England and Wales*, 1862, 12mo, 42 maps.
Dawson (R. K.). *Plans of the Cities and Boroughs of England and Wales*, 1832, folio, 277 plates.
Dix (T.) and W. Darton. *Complete Atlas of the English Counties*, 1822, folio, 42 maps.
     Other editions 1830, 1833, 1848.
Dodsley (R.) and J. Cowley. *Geography of England*, 1744, 8vo, 55 maps.
—— *New Sett of Pocket Maps of all the Counties of England and Wales*, 4to, 1745 (Re-issue of preceding).
Drayton (Michael). *Polyolbion*, 1612, 18 maps (40).
     Re-issue 1613, 1622 (in this edition second part added with 12 extra maps, 30 in all). Reproduction in 1890 by Spencer Society.
Dugdale (J.). *New British Traveller*, 1819, 4to, 45 maps.
Dugdale (T.). *Curiosities of Great Britain*, 1835, 8vo.
     Other editions 1843, 1848 (58 maps by J. Archer), 1854–60.
Duncan (J.). *Complete County Atlas of England and Wales*, 1833, folio, 44 maps.
     Other editions 1837, 1838, 1840, 1845, 1858, 1865.
Ellis (J.). *Ellis's English Atlas*, 1766, oblong 8vo, Robert Sayer and Carington Bowles, 48 maps (50 in list, but Nos. 3 and 4 not issued).
     Other editions; 1766, Carington Bowles and Robert Sayer, 54 maps (Channel Is. being added): 1768, Robert Sayer, Thomas Jefferys and A. Dury: 1773, Robert Sayer: 1777, R. Sayer and J. Bennett.
Ellis (G.). *Ellis's New and Correct Atlas of England and Wales*, 1819, 4to, 44 maps.
Fisher, Son & Co. *County Atlas of England and Wales*, 1842–5, 4to, 43 maps.
Fullarton & Co. (A.). *Parliamentary Gazetteer of England and Wales*, 1840, 8vo.
     Other editions 1843, 1845, 1846, 1848.
Gardner (T.). *Pocket Guide to English Traveller*, 1719, sm. 4to, 100 road maps reduced from Ogilby.
Gibson (J.). *Universal Magazine*, 1763–4, 1765–9, 12 maps.
—— *New and Accurate Maps of the Counties of England and Wales*, 1759, 12mo, 53 maps.
     Other editions 1770, 1779.
Gorton (J.) and S. Hall. *Topographical Dictionary of Great Britain and Ireland*, 1833, 8vo, 54 maps.
Gray (G. C.). *New Book of the Roads*, 1824, 12mo, 49 maps.
Green (W.). *Picture of England*, 1804, 2 vols., 8vo.
Greenwood (C. & J.). *Atlas of the Counties of England*, 1834, folio, 46 maps.
Grose (F.) and John Seller. *Supplement to the Antiquities of England and Wales*, 1777–87, sm. folio, 52 maps.

Hall (S.). *New British Atlas*, 1833, 4to, 54 maps.
  Other editions 1834, 1836.
—— *A Travelling County Atlas*, 1842, 8vo.
  Other editions 1843, 1845 (46 maps), 1846, 1848, 1850, 1852, 1854, 1857, 1866, 1873, 1875, 1880, 1885.
—— *New County Atlas*, 1847, 4to.
—— *The English Counties* [1860], folio, 41 maps.
Harrison (J.). *Maps of the English Counties*, 1791, folio, 38 maps.
  Another edition, 1792. Reissued as *General and County Atlas*, 1815.
Hinton (J.). *Universal Magazine*, 1765–73, 39 road maps after Ogilby.
Hobson (——) and J. and C. Walker. *Hobson's Fox Hunting Atlas*, 1850, folio, 42 maps.
  Re-issued 1866, 1868, 1870, 1872, 1880.
Hodgson (O.). *Pocket Tourist and English Atlas*, 1820, 12mo, 43 maps.
Hughes (W.). *National Gazetteer*, 1868, 8vo, 68 maps.
Jansson (J.). *Newer Atlas*, 1638, folio, 56 maps.
  Other editions; 1638 (Dutch text), 1642 (Dutch text), 1646 (Latin text), 1646 (French text), 1647 (French text), 1647 (German text), 1649 (German text), 1652 (French text), 1652 (Dutch text), 1656 (French text), 1659 (Latin text), 1666 (Latin text), 1683 (publishers Schenk and Valk), 1710 (Allard), 1714 (D. Mortier), 1724 (Schenk and Valk).
Jefferys (T.). See Kitchin (T.).
Jenner (T.). *Direction for the English Traveller*, 1643 (49).
  Other editions 1657, 1662, 1668, 1676, 1677.
Johnson (T.). *Johnson's Atlas of England*, 1847, 4to, 41 maps.
Keer (Peter van den). (Collection of Maps of the Counties of England and Wales). No title 1599, oblong 8vo, 37 maps.
  Re-issued 1617, 46 maps, 1627 (taken over by Humble and maps increased to 63): 1627, 1630, 1646, 1662, 1666, 1676.
Kelly & Co. *Post Office Directory Atlas of England and Wales*, 1860, 4to, 45 maps.
Kitchin (Thomas). *London Magazine*, 1747–60, 8vo, 56 maps.
—— and T. Jefferys. *Small English Atlas*, 1749, 4to, 50 maps.
  Other editions 1751, 1775, 1785 and 1787 (new title; *An English Atlas or Concise View of England and Wales*).
—— *England Illustrated*, 1764, 4to, 54 maps.
—— *Kitchin's English Atlas*, 1770, 4to, 54 maps.
—— and Eman. Bowen. *The English Atlas*, 1765, 4to, 47 maps.
—— *Kitchin's Pocket Atlas*, 1769, oblong 8vo, 47 maps.
Langley (E.) and Belch. *New County Atlas*, 1818, 4to, 53 maps.
  Another edition 1820 (61).
Laurie and Whittle. *New and Improved English Atlas*, 4to, 1807.
—— *New Traveller's Companion*, 1810, 8vo, 26 maps.
Leigh (S.). *New Atlas of England and Wales*, 1820, 12mo, 56 maps.
  Other editions 1825, 1826, 1830, 1831, 1833, 1835, 1837, 1839, 1840, 1842, 1843 (Leigh's *New Pocket Atlas of England and Wales*).
Lewis (William). *New Travellers' Guide*, 1819, 12mo.
  Other editions 1835, 1836.
Lewis (S.). *Topographical Dictionary of England*, 1831, 4to, 43 maps.
  Other editions 1833, 1835, 1837, 1840, 1842, 1844, 1845, 1848, 1849.
—— *Topographical Dictionary of Wales*, 1833, 4to, 12 maps.
  Drawn by Creighton, engraved by Walker.
  Other editions 1840, 1843, 1849.
Lodge (J.). *Political Magazine*, 1782–90, 8vo, 54 maps.
  Maps re-issued in volume form in 1795.
London Magazine. See Kitchin (T.).
Luffman (J.). *New Pocket Atlas and Geography of England and Wales*, 1803, 8vo, 54 circular maps (51).
  Other editions 1805, 1806.

Martin's Sportsman's Almanack, 1819, 8vo.

M'Leod (W.). *Physical Atlas of Great Britain,* 1861, 8vo, 30 maps.

Miller (R.). *Miller's New Miniature Atlas,* 1810, 12mo, 56 maps.
  Other editions 1820 (title changed *Darton's New Miniature Atlas*), 1825.

Moll (Herman). *New Description of England and Wales,* 1724, folio, 50 maps (41).
—— Re-issued 1724 (with new title *A Set of Fifty New and Correct Maps of England and Wales*),
  1728 (*New Description of England and Wales*), 1739, 1747, and 1753.

Morden (R.). *Pocket Book of all the Counties of England and Wales* (1680), 8vo, 52 maps.
—— Another edition 1750 (*A Brief Description of England and Wales*), 12mo, printed for
  H. Turpin.
—— *Camden's Britannia,* 1695, folio, 50 maps engraved Sutton Nicholls and John Sturt (54).
  A few copies were printed on large and thick paper.
—— Other editions 1715, 1722 (in 2 vols. with 2 maps of Scotland, North and South, instead of one
  general map; some of the maps have considerable corrections and additions), 1737, 1753 and
  1772.
—— *New Description and State of England,* 1701, 8vo, 54 maps.
—— Other editions 1704, 1708 (new title, *Fifty-six New and Accurate Maps of Great Britain . . .
  corrected and enlarged by H. Moll*), 1720–31 (new title, *Magna Britannia et Hibernia,* and 5
  new maps), 1738.

Moule (T.). *The English Counties Delineated,* 1836, 4to (63).
  Other editions 1837 (58 maps and plans), 1838, (1841) (with new title, *Barclay's Complete
  and Universal Dictionary*), 1842, 1848, 1850 and 1852.

Murray (T. L.). *Atlas of the English Counties,* 1830, folio, 44 maps.
  Other editions 1831, 1832, 1834.

Nightingale (J.). *English Topography,* 1816, 4to, 56 maps.
  Another edition 1820.

Ogilby (J.). *Britannia . . . Description of the Principal Roads Thereof,* folio, 1675, 100 road maps (53).
  The first survey of the roads of England and Wales. Some copies were printed on large
  and thick paper, red ruled.
  There were two issues of the first edition; the earliest has 7 leaves of text on the history of
  London, in the later edition this text is condensed to 4 leaves. A third edition was issued in
  the same year without the accompanying text.
  Reprinted 1698 from the same plates.
  Reduced facsimile copy in colour (oblong folio), 1939.

Osborne (T.). *Geographia Magnae Britanniae,* 1748, 8vo, 63 maps.
  Another edition 1750.

Owen (J.) and E. Bowen. *Britannia Depicta,* 8vo, 1720 (56).
  Other editions 1721, 1723, 1724, 1730, 1731, 1734, 1736, 1749, 1751, 1755, 1759, 1764.

Perrot (A. M.). *L'Angleterre ou Description Hist. et Topographique du Royaume Uni de la Grand Bretagne
  par G. B. Deeping,* 1823, 6 vols., 12mo, 75 maps and views.
  Other editions 1828 (2 editions) and 1835.

Pigot & Co. (J.). *British Atlas of the Counties of England,* 1829, folio, 41 maps (62).
  Other editions 1831, 1832, 1833, 1838, 1839, 1840, 1844 (1846 taken over by Slater).
—— *London and Provincial New Directory,* 1827, 8vo.
  Other editions 1829, 1832.
—— *Pocket Topography and Gazetteer of England,* 1835, 8vo, 40 maps.
  Another edition 1842.

Pinnock (W.). *History and Topography of England and Wales,* 1825, 12mo, 42 maps.

Ramble (Reuben). *Travels through the Counties of England,* 1845, 4to, 40 maps.

Reid (W. H.) and J. Wallis. *The Panorama or Travellers' Instructive Guide,* 1820, 8vo, 53 maps.

Reynolds (J.). *Travelling Atlas of England,* 1848, 8vo, 32 maps.
—— *Geological Atlas of Great Britain,* 1860, 8vo, 32 maps.
  Other editions 1864, 1889.
—— *Portable Atlas of England and Wales,* 1864, 8vo, 32 maps.

Robins (J.). *Atlas of England and Wales,* 1819, 4to, 45 maps.

Rocque (J.). *English Traveller*, 1746, 8vo, 3 vols.
—— *Small British Atlas*, 1753, sm. 4to, 54 maps.
    Other editions 1762, 1764, 1769 (with new title *England Displayed*, 29 maps).
Rodwell (M. M.). *Geography of the British Isles*, 1834, 8vo, 59 plates.
Rowe (R.). *English Atlas*, 1816, folio, 46 maps.
    Issued from 1829 onwards as *Teesdale's New British Atlas*.
Russell (P.) and Owen Price. *England Displayed*, 1769, 2 vols., folio, 50 maps.
Saxton (Christopher). (*Atlas of England and Wales*), 1579, 35 maps (38).
    Re-issued 1645 by William Webb: 1665, 1689 and 1693 by Philip Lea: 1720 (Willdey): 1749 (T. Jefferys). See also Camden's *Britannia*.
Sayer (R.). *English Atlas or Concise View of England and Wales*, 1787, 4to, 48 maps.
Seller (John). *Anglia Contracta* (1695), 8vo., 66 maps.
    Other editions 1696: 1701 (Camden's *Britannia* abridged, J. Wild), 2 vols., 8vo, 59 maps: 1703, 8vo, 66 maps: reprinted 1787 in Grose's *Antiquities* without acknowledgment.
Senex (J.). *Actual Survey of all the Principal Roads of England and Wales*, 1719, oblong 8vo, 100 maps reduced from Ogilby.
    Other editions 1742, 1757 (*Roads through England Delineated*, printed John Bowles), 1767 (Kitchin's *Post-Chaise Companion*), 1770 French edition, 1767 (*Nouvel Atlas d'Angleterre*, Paris, Desnos), 1775 (Jeffery's *Itinerary*, printed Sayer and Bennett), and 1775.
Simons (Matthew). *Direction for English Traviller*, 1635, sm. 4to, 39 maps.
    Re-issued 1636, 1643, 1657 (published Thomas Jenner), 1662, 1668, 1677 (published John Garrett).
Simpson (S.). *Agreeable Historian or Compleat English Traveller*, 1746, 8vo, 41 maps.
Slater (I.). *Slater's New British Atlas*, 1846, folio.
    Other editions 1847, 1848, 1855, 1857, 1859.
Smith (C.). *Smith's New English Atlas*, 1804, folio, 42 maps.
    Other editions 1808, 1818, 1821, 1827, 1834, 1837, 1841, 1848 and 1864.
—— *New English Atlas*, 1822, 4to, 44 maps.
—— *New Pocket Companion to the Roads of England and Wales*, 1826, 8vo, 126 strip maps.
Speed (J.). *Theatre of the Empire of Great Britain*, 1611, folio, 67 maps (36 and 47).
    Trial issue 1605–10.
    Other editions 1612, 1614, 1616 (all published Sudbury and Humble), 1627 and 1631 (George Humble), 1646 and 1650 (William Humble), 1662 (Roger Rea), 1676 (Bassett and Chiswell, issued in two ways, with text on back of maps and with plain backs), 1680 (J. Seller), 1696, 1710, 1713 and 1743 (Overton), 1770 (Dicey and Co.).
Taylor (T.). *England Exactly Described* (1715), 8vo, 42 maps.
    See also Blome (R.).
    Other editions 1716, 1718, 1731 (Tho. Bakewell).
—— *Principality of Wales exactly Described*, 1718.
Teesdale (H.). *New British Atlas*, 1829, 4to, 45 maps.
    Other editions 1830, 1831, 1832, 1833, 1835, 1840, 1848 (taken over by H. G. Collins).
—— *New Travelling Atlas*, 1830, 4to, 45 maps.
—— Another edition 1842, 1848 (new title, *Travelling Atlas of England and Wales*), 1850, 1852, 1860.
Tymms (S.). *Family Topographer*, 1832–43, 8vo, 32 maps.
Universal Magazine, 1747–66, 8vo, 51 maps.
    Maps by Bowen, Kitchin and Seale, published J. Hinton.
Walker (J. & C.). *British Atlas*, 1837, folio, 47 maps.
    Other editions 1837, 1838, 1839, 1840, 1841, 1842, 1843, 1844, 1846, 1849, 1852, 1861, 1862, 1865, 1870, 1873, 1879.
Wallis (J.). *New Pocket Edition of the English Counties*, 1810. 12mo, 44 maps.
    Another edition 1814.
—— *New and Improved County Atlas*, 1812, folio, 42 maps.
    Another edition 1813.
Walpole (G. A.). *New British Traveller*, 1784, folio (23 plates, some containing 2 or 3 maps to a plate).
    Another edition, 1794.

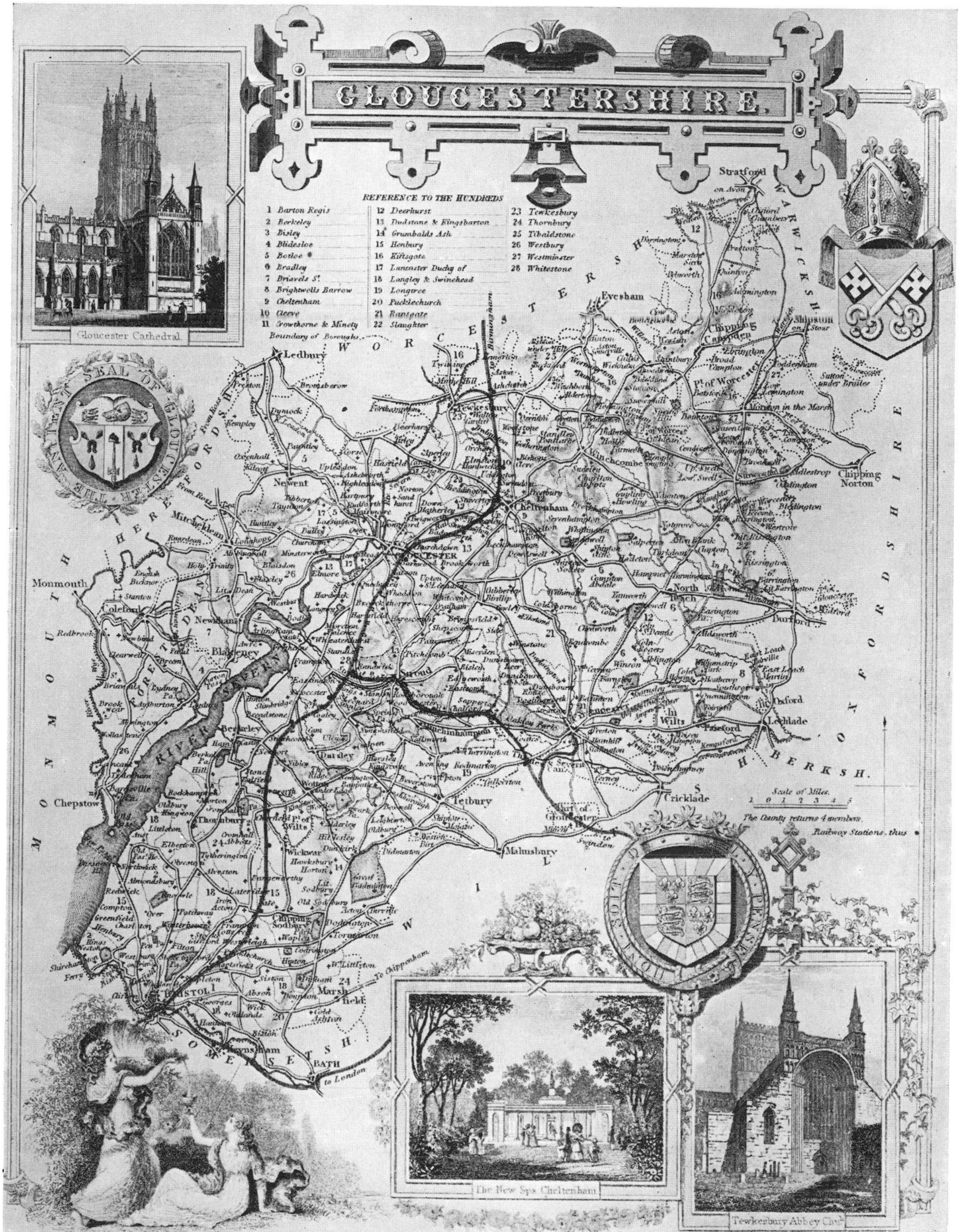
GLOUCESTERSHIRE.
REFERENCE TO THE HUNDREDS
1 Barton Regis
2 Berkeley
3 Bisley
4 Bledisloe
5 Botloe
6 Bradley
7 Briavels St.
8 Brightwells Barrow
9 Cheltenham
10 Cleeve
11 Crowthorne & Minety
12 Deerhurst
13 Dudstone & Kingsbarton
14 Grumbalds Ash
15 Henbury
16 Kiftsgate
17 Lancaster Duchy of
18 Langley & Swinehead
19 Longtree
20 Pucklechurch
21 Rapsgate
22 Slaughter
23 Tewkesbury
24 Thornbury
25 Tibaldstone
26 Westbury
27 Westminster
28 Whitestone
Boundary of Boroughs
Gloucester Cathedral
The New Spa Cheltenham
Tewkesbury Abbey Chur
The County returns 4 members.
Railway Stations, thus
Scale of Miles.

63 T. MOULE'S *GLOUCESTERSHIRE* (8″ × 10″)
from *The English Counties Delineated*, 1836

64   DETAIL FROM JOHN ROCQUE'S 2″ MAP OF BERKSHIRE, 1761

Weekly Dispatch, 1858–62, folio, maps engraved by Davies, Dower and Weller.
—— *The Dispatch Atlas*, 1863, folio, 75 maps.
      Other editions 1864, 1876 (taken over by G. W. Bacon).
Whitaker (G. & W. B.). *Travellers' Pocket Atlas*, 1823, 8vo, 43 maps.
Wilkes (J.). *Encyclopædia Londinensis*, 1810–28, 4to, 40 maps.
Wyld. *Atlas of English Counties*, 1842, 4to.

CHAPTER IX

Scotland and Ireland

THE printed cartography of Scotland starts with Ptolemy's representation of Great Britain. In the first and subsequent editions (see Chapter I) up to the edition of 1508 Scotland is delineated twisted over to the right. This was first corrected in the edition of Bernardus Sylvanus of 1511, Scotland after that date being given an upright position.

Preceding and contemporary with the first printed maps of Ptolemy, portolan or sea charts were constructed. These portolans include the whole of the British Isles, of which Scotland forms the weakest feature. The earliest examples are very rudimentary, with the north not even filled in. Though limited in scope, they form an attempt independent of the classics to map these regions. The subject has been very ably discussed by Mr. M. C. Andrews (see Authorities), who divides the 14th- and 15th-century examples into four groups: (1) Genoese; (2) Venetian; (3) Catalan, and (4) Late 15th century. Further variations occur in the 16th century, but towards the middle of this period they were superseded by the more detailed maps of the land geographers.

In 1540 James V of Scotland set out on a tour of his kingdom, accompanied by Alexander Lindsay, pilot to the expedition. Observations were taken, and probably a map compiled, but if so this has been lost. This event undoubtedly forms a landmark in the development of the national topography, and Lindsay's work may possibly have supplied a basis for the Scottish portion of the next important map to appear. This was Lily's map of the British Isles, Rome, copper-plate, 1546. George Lily was a member of the household of the exiled Cardinal Pole. Other English exiles probably contributed to the formation of this map. As far as it concerns Scotland, Lily's map is much superior to any previous-known representation. It is rare, only seven examples being known to date (including Paris, London and Dublin). Lily's map was popular, and re-issued as follows:

1549   Antwerp (B.N., Paris).
1555   London (B.N., Paris).
1556   Rome (21 known examples).
1556   Venice Vavassor (B.M.).
1558   Rome Sebastianus a Regibus Clodiensis (23 known examples).
1562   Venice Bertelli (12 known examples).
1563   Venice Camotius (3 known examples).
1589   Rome, Marc. Clod. (3 known examples).

They were all printed from copper-plates, except the Antwerp and the Vavassor editions. The first two and the 1558 and 1589 editions have the north on the right of the map. They all differ in engraving and size with minor omissions and additions. Lily's engraved map was even incorporated by the portolan makers in their work, Agnese and Millo among others.

About the year 1550 John Elder of Caithness is said to have compiled a map of Scotland, but the manuscript is now lost.

The earliest-known separately printed map of Scotland is an anonymous map printed in Italy about 1568. It is in effect based on the 1558 map of Sebastianus a Regibus Clodiensis and so a derivative of Lily. Only two examples are known.

In 1564 a map of importance was printed. This was the great map of the British Isles published by Mercator, engraved on eight copper-plates on a scale of $14\frac{1}{2}$ miles to the inch, with the title "Angliae Scotiae & Hiberni noua descriptio 1564." Three examples are so far known to survive, in Breslau, Rome and Perugia. A reproduction was issued in 1891. Mercator's map was copied both in whole and in part, on a reduced scale, by various geographers. These reductions were not, however, exact copies, being amended as information became available.

Ortelius's map of Scotland first appeared in the 1573 additamentum, and was incorporated in his atlas of that date and in subsequent editions (see Chapter V). It is based, with modifications, on Mercator. The north is to the right of the map.

Mercator's single-sheet map of Scotland first appeared in the 1595 edition of his atlas. Based on the 1564 map of the British Isles, it is corrected in several places. Later editions were issued by Hondius (see Chapter V). Mercator's conception was also followed by Hole and by Blome.

Another notable map is John Leslie's map of Scotland. John Leslie, Bishop of Ross, the defender of Mary, Queen of Scots, was born in 1527 and died in 1596. Exiled in 1574, he stayed first in France, later in Rome, where he occupied his leisure in writing his Latin *History of Scotland*, which was published in 1578. In 1579 he was appointed suffragan and vicar-general of the diocese of Rouen.

His maps are as follows:

1578. Rome, $21\frac{1}{2}$ by $16\frac{1}{2}$ inches. Natalis Bonifacius. (No copy known.)
1578. Rome, $7\frac{1}{2}$ by 11 inches.
(1586.) Rouen, 15 by $20\frac{1}{2}$ inches. Map still dated 1578.
(1590–96?.) Anon. No author, printer or date.

The first of these is extremely rare. There is no copy in the British Museum or in any of the Scottish Libraries, and its absence has caused considerable variation of opinion on the subject.

The two Rome maps are entirely different. The smaller is based on an older model, viz. on Lily's map of the British Isles, with a close cluster of islands for the Orkneys and a similar cluster for the Western Isles. It was issued in *De Rebus Gestis Scotorum*.

The large map of Leslie, on the other hand, resembles the Ortelius variation of Mercator, that is, a far better outline for the kingdom generally, and the outer isles are inserted. But it departs from Ortelius, in that inner isles preserve the older conception, and the large island of Hirta is retained, though its position is changed. The Rouen edition is a re-issue of the larger Rome edition, and the anonymous map a copy, but with all reference to its provenance removed and a new title substituted. Of the last three editions several examples are known.

A few years later, in 1583, an entirely new map was produced in Paris by Nicolas de Nicolay. Born in 1517 at La Grave d'Oissans in Dauphiny, he at first adopted a military career, taking part in the siege of Perpignan in 1542. After that date he set out on his travels. He travelled for sixteen years on various missions in Germany, Scandinavia, England and Spain. During this period, namely, in 1546–7, he visited England, where he was hospitably received by the Admiral Dudley Earl of Northumberland, and undertook a voyage with the latter in Scottish coastal waters. Possibly due to his travels, it was not till some years later that Nicolay's map of Scotland appeared in print. His map of 1583 gives an excellent representation of Scotland for that time, and he must have had the assistance of some Scottish survey to help him. This cannot now be traced, but may have been Lindsay the pilot's observations, for Nicolay's map is a marine not a land map, and he was close enough in time to James V's tour for Lindsay's work to have been in circulation. At the end of the century Braun and Hogenberg issued an excellent plan of Edinburgh (14).

The 17th century opens with the small but attractive map of Scotland in Camden's *Britannia*. This was first issued with Latin text in 1607 and re-issued in 1610 and 1637 without text on the reverse. It was engraved by William Hole, one of the earliest of English engravers.

Hole's map was followed by one of the most decorative and certainly the most popular of all early maps of Scotland, John Speed's "The Kingdome of Scotland" (65). This map was superbly engraved, and contains side borders of full-length portraits of the reigning Stuart family. There were many re-issues of Speed's maps (see Chapter VII), but the impressions became fainter, and for the 1653 edition the plate was re-engraved, the Stuarts were removed and four figures showing national costume inserted. These were of much cruder workmanship.

In the next few years Blaeu, Jansson and Hondius all issued maps of Scotland in their respective atlases (see Chapter V). Though they show little progress geographically, they are all extremely attractive in the decorative sense, especially when in contemporary colouring, and well worthy of a place in any collection. Hollar engraved a map of England, Ireland, Scotland and Wales at the time of the civil war, and this went into several editions.

But the most important event in the 17th century was the issue of the first national atlas of Scotland. This was published by W. and J. Blaeu in 1654. It formed volume V of their *Atlas Novus*, but is complete in itself, and was sold as a separate entity. It contains a finely engraved title-page and 49 maps (including 3 general maps and 46 maps of the provinces and islands). Of the regional maps, no less than 36 were based on the surveys of Timothy Pont, 3 by Robert Gordon of Straloch, 1 by his son James Gordon, and 6 for which no attribution is given. Timothy Pont was a minister born about 1565. Only one of his maps is dated, viz. 1596. He was unable to find a publisher for his manuscripts, in fact they were hardly suitable in their original state, and credit is due to Gordon for the skill with which he redrew the maps in a uniform manner suitable for the press. Blaeu's *Atlas of Scotland* went into several editions, being issued with Dutch and with Latin text in 1654, with Latin text in 1662, French text in 1663 and 1667, Dutch text in 1664 and Spanish text in 1672. As regards the general map of Scotland, the Gordon-Blaeu outline set a new standard that was followed by other cartographers for about a hundred years.

Jansson, the rival of Blaeu, did not publish an atlas of Scotland, but included some maps of Scotland in his English volume. Jansson's maps were:

General map of Scotland.
Lothian.
Islands.
Orkney and Shetlands.

These appear in the Latin edition of his atlas of 1659. Another issue contains 7 maps, namely, General Map, Northern Scotland, Central (Tay to Murray Firth), Southern (Tay to Border), Lothians, Lewis and Harris, Inner Isles, etc. All these were re-issued by Valk and Schenk towards the end of the century.

Further maps of Scotland from the Dutch presses in the 17th century are those of Van der Aa, Allard, Danckerts, De Wit and Visscher (see Chapter V). English maps of Scotland during this period are represented by Blome 1673, Green 1689 and Morden 1695, and in this

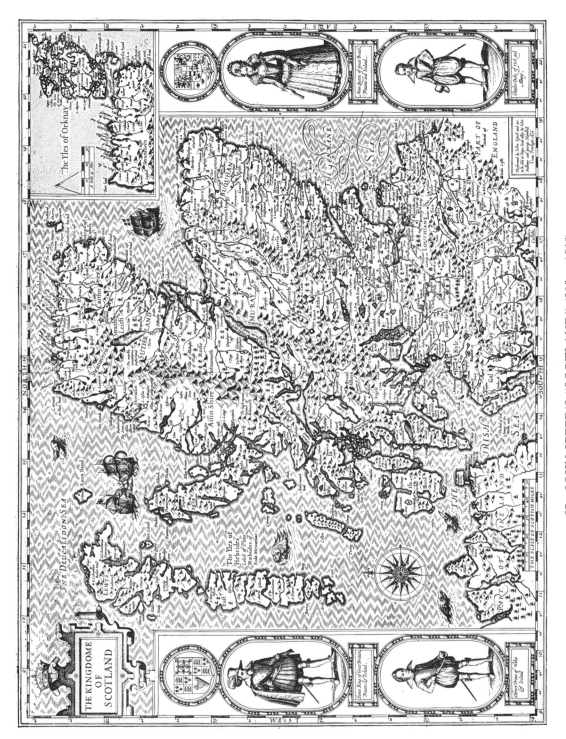

65 JOHN SPEED'S "SCOTLAND" (20" × 15¼")
from *The Theatre of the Empire of Great Britain*, 1611

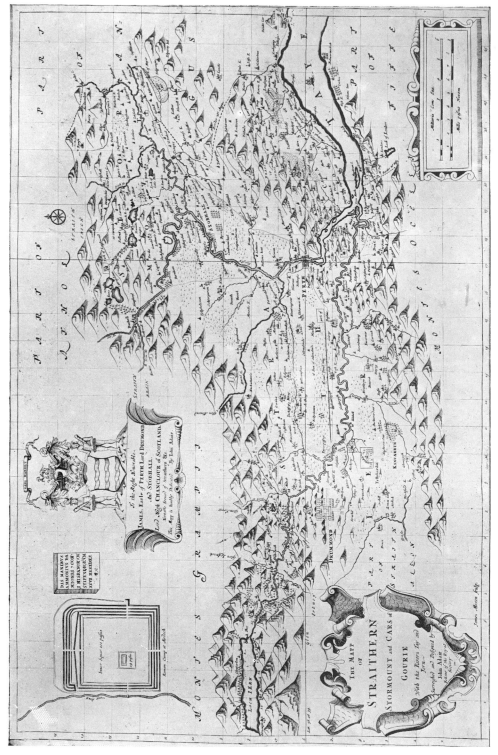

66　JOHN ADAIR'S MAP OF STRAITHERN, c. 1688 (26″ × 17″)

last year, 1695, a large Italian map by Coronelli showing Scotland on two sheets was printed. From the French school came handsome maps by Sanson and Jaillot (see Chapter VI).

To go back a little in time in the last quarter of the 16th century, printed charts began to appear. The maps from these marine atlases are far more scarce than their land counterparts. They commence with the atlas of Waghenaer in 1585, and continue through the 17th century in the sea atlases of Blaeu, Colom, Jacobsz, Jansson, Goos, Doncker, Van Loon, Robijn and Keulen (see Chapter V).

Three English hydrographers produced atlases in the 17th century. The first to appear was Robert Dudley's *Arcano del Mare*, printed in Florence in 1646, which includes charts of the coast of Scotland.

In 1671–2 John Seller published his *English Pilot*. The following charts relate to Scotland:

North Sea.
East Coast of Scotland.
Shetland and Hebrides.
Great Britain.
East side of Ireland (including part of Scotland).

And in 1693 Capt. Greenvile Collins issued *Great Britain's Coasting Pilot*, giving greater detail in the following charts:

(Anglesea to Mull of Galloway.)
East Coast of Scotland, Orkney and Shetland.
(Edinburgh Firth.)
(Leath, with inset view.)
Fifeness to Montrose, inset Aberdeen.
(Firth of Murray.)
Islands of Orkney.
Chief harbours in the Islands of Orkney.
(Part of the Main Island of Shetland.)
South Part of the Isles of Shetland.

There were many editions of Collins' *Atlas* (see page 62).

Charts of the Scottish coast also appear in the *Neptune François*, published by H. Jaillot.

The first important publication in the 18th century was likewise a marine atlas, John Adair's *Description of the Sea Coasts and Islands of Scotland*, Part I (all published), 1703. It comprises title and 18 pages of text and 6 charts:

General Map (after Nicolay).
Sunderland Point to St. Abb's Head.
Firth of Forth.
Firth and River of Tay.
Town and Water of Montrose.
Red Head of Aberdeen.

A real improvement was made in mapping Scotland by Herman Moll, a Dutchman resident in England. His map of the *North part of Britain called Scotland* was printed on two large sheets (usually joined together), with large vignettes down each side of views of the principal towns taken from Slezer. There are the following issues:

Printed for H. Moll . . . D. Midwinter . . . and Tho. Bowles, 1714.
Printed for H. Moll . . . I. Bowles . . . and Tho. Bowles, 1714.
Printed for H. Moll . . . I. Bowles . . . Tho. Bowles and John King, 1714.

Similar large-scale maps were issued by John Senex and Henry Overton, based on Sanson. A few years later an anonymous map appeared copied from Moll, but with the title changed to *A New Map of Scotland or North Britain*, engraved with straight lines between the principal towns to show distances. There are apparently three versions of this: one dated 1731, sold by T. Bowles and John Bowles; another 1732 (see *Early Maps of Scotland*, 1936); and one Printed and Sold by William Knight Delineated and Engraven by Sutton Nicholls. From 1725 General Wade started making his roads from Inverness to the south, and from this time many military surveys were made, though these were not directly printed.

In 1725 the second atlas of Scotland appeared. This was by Herman Moll, under the title *A Set of 36 New and Correct Maps of Scotland divided into Shires*, etc., oblong 4to in size; re-issued in 1745 and reprinted 1896.

The rising of 1745 still further stimulated map production, and several publications appeared about this date: Elphinstone 1745, Overton (a re-issue of the 1715 map but with roads added), Bowen 1746 among others, and the Great Military Survey was started in 1747.

In 1749 the first pocket atlas of Scotland appeared: this was compiled by Thomas Kitchin, and entitled *Geographia Scotiae*. It contains 33 maps.

A year later James Dorret brought out a greatly improved map of Scotland on four sheets, 1750. These were re-issued by Dorret on a smaller scale in 1751 and 1761. Dorret's map continued the standard map for about thirty years, till it was in part superseded by John Ainslie's map of 1789. Intervening map-makers were Bowles, Kitchin, Dunn, Armstrong and Cary.

In 1777 Mostyn John Armstrong produced a small 4to atlas of the county maps of Scotland. This neat little atlas contains 30 maps, the first edition being printed by Sayer and Bennett. It was re-issued in 1787 with Sayer's name only, and again in 1794 with the imprint of Laurie and Whittle.

Ainslie's map on nine sheets on a scale of 4 miles to the inch, 1789, held the field till the production of Greenwood's map.

About the turn of the half-century some large-scale regional surveying was begun. Colonel Watson, on the command of the Duke of Cumberland, began his military survey of the Highlands in 1747. The work was placed under the charge of Lieutenant Roy, and 84 sheets were completed by 1755, the work having been gradually extended to the Lowlands. These military surveys, precursors of the Ordnance Survey, were not published, and apparently were not made available for civilian use till the time of Aaron Arrowsmith. Large-scale work was nevertheless undertaken by private surveyors, of which the following are some examples:

Mackenzie (Murdoch). Orkneys, 8 sheets, 1750, 1767, 1776, 1791.
Ainslie (John). Forfar, 4 sheets, 1794.
—— Kirkcudbright, 4 sheets, 1797.
—— Renfrew, 4 sheets, 1796.
—— Wigtown, 2 sheets, 1782.
Armstrong (Capt.) and Son. Ayrshire, 6 sheets, 1773.
—— Berwick, 4 sheets, 1770.
—— Lothians, 6 sheets, 1773.
—— Peebles or Tweedale, 2 sheets, 1775.
Laurie (J.). Midlothian, 4 sheets, 1763.
Ross (C.). Dumbarton, 2 sheets, 1777.
—— Stirling, 1780.
Stobie (J.). Perth and Clackmannan, 9 sheets, 1783.
—— (M.). Roxburghshire, 4 sheets, 1770.

All the above are difficult to come by.

During the course of this century several maps of Scotland were produced abroad: in France, maps by Sanson, Nolin, De Fer, Le Rouge, Robert, Bellin, Tardieu, Brion and

Palairet; in Germany and Austria, Homann, Seutter, Lotter and Reilly and Schraembl, and in Italy, Zatta.

There is thus a wealth of material in the 18th century to draw upon, only a portion of which has been cited. The most important maps are those by Adair (66), Moll, Dorret and Ainslie.

In 1800 Thomas Brown issued an atlas of Scotland in 4to, with 26 county maps, and in 1801 Cary printed a fine 4-sheet map of the whole of Scotland. An even larger map on 12 sheets was published by John Stockdale in 1806. The finest map of this period was, however, compiled by Aaron Arrowsmith on a scale of 4 miles to the inch. This was printed on 4 sheets, 1807. Arrowsmith was the first to be allowed to draw on the military survey hitherto unpublished, and thus to a point anticipated and vied with the early work of the Ordnance Survey, which had already commenced to print its maps of southern England at this date.

Single-sheet maps of this period are too numerous to be listed in a book of this nature, but mention must be made of William Forrest's Survey of Lanark on 8 sheets in 1816; Grassom's Stirling; James Robertson's Aberdeen, 6 sheets, 1822; Sharp, Greenwood and Fowler's Fife and Kinross, 4 sheets, 1828. The complete atlases are as follows: Lothian's *County Atlas of Scotland* appeared in 1827 with 34 maps, and went into several editions. It is 4to size. The finest county atlas of Scotland in the 19th century was that by John Thomson & Co., Edinburgh, first in 1831 and again in 1832. This folio production was issued in parts, and contains 2 folding plates and 30 maps on 58 sheets. An unusual and interesting feature of this atlas is that names of surveyors and other persons are given on the maps attesting to their accuracy. A corrected edition was issued in 1869. Later atlases of the 19th century were issued in 1838 by William Blackwood, engraved by Lizars, in 4to, republished in 1839 and 1845, and again in 8vo format in 1853; Black's county atlas of Scotland, 1848; Wilson's *Gazetteer*, 1854–7; and Phillips' *Atlas*, 1858 and 1860.

In the history of maps, Ireland is of particular interest as the most westerly point known to the ancients. The earliest mapping of Ireland of which there is a record is that given by Ptolemy in his map of the British Isles, first printed in Bologna in 1477 (see Chapter I). Ptolemy's Ireland is distorted and out of place, lying too far north relatively to Britain, but it depicts 15 rivers, 5 promontories, 11 towns and 9 islands, a remarkable feat for this era. It was far less satisfactorily represented in the Mediæval period, when it is usually shown in ideographic form as part of a general map. There is no ancient Irish manuscript extant comparable to the earliest English examples, which is somewhat surprising, considering the high state of culture of Ireland in early times, when Irish thought and enterprise and Irish monks spread to England and the Continent. Ireland is shown of course in the more elaborate mappaemundi with some detail, as in the Hereford MS. (see Chapter II), where Dublin, Bangor and Armagh are marked.

But the first stirrings of interest in the mapping of Ireland after the days of Ptolemy begin with the marine charts or portolans dating from the 14th century onwards, such as Vesconte, Dalorto and Sanuto. Andrews divides these charts into six types or models, those of Benincasa 1468, Anon 1469, Venetian Early 15th century, Homen 1558, Martines 1583 and Olives 1614. From the 13th century onwards the Italians, particularly those from Florence and Lucca, had a flourishing direct trade with Ireland. For example, the Friscobaldi had representatives in Ireland in 1282, the Donati held the dues of Dublin in 1284, and also had establishments in Cork and Waterford, and the Ricardi's sphere of influence was over the whole of southern Ireland. There was an Irish hospital in Genoa about 1160. Later, in the early years of the 15th century, the markets of Brabant were visited by Irish merchants, where contact was again made with persons interested in cartography. Hence the comparative superiority of the Italian and Catalan portolan charts. These hand-drawn charts served as the basis of the printed

Tabula Nova in the later editions of Ptolemy, that is, from the 1513 Strassburg edition onwards.

The first "modern" printed maps of Ireland likewise had their beginning in Italy, Ireland being shown on Lily's map of the British Isles of 1546 and its 8 derivatives (see the beginning of this chapter). The Irish portion of this map is based neither on Ptolemy nor the Portolans. Only 3 rivers and 15 towns are named, and this land map is poor in comparison with the sea charts, with their 60 or more names.

In addition to the maps of Great Britain, however, Ireland is represented in four separate anonymous maps to be found in Lafreri and similar collections. These are as follows:

Hibernia sive Irlanda (Rome 1560). The south is at top of map.
Hibernia insula . . . 1565. Venice, Zalterius, 1566.
Hybernia nunc Irlant (Venice 1565).
Hybernia nunc Irlant. Rome, Duchetti.

The first of these is the oldest-known printed map of Ireland. The map of Ireland published by Porcacchi in his *Isolario* of 1572 and subsequent editions is based on the later of these. In the same year a map of Ireland was issued by Camocio.

A few manuscript maps by natives or residents of this period have survived, the earliest being one compiled for Henry VIII. It is anonymous. A finer map was produced by John Goghe: it is signed and dated 1567. This at one time belonged to Cecil, and bears his annotations. A reproduction was printed in 1834. Laurence Nowell drew two maps of Ireland about 1576: these 2 maps were reproduced about 1860 by the Ordnance Office. Robert Lythe, John Browne and Francis Jobson were also surveying during this period, the first of whom compiled a general, and the others provincial, maps. Baptista Boazio, an Italian, drew a fine map of the whole of Ireland; it is undated, but was probably finished in 1586–8. This manuscript is preserved in Trinity College, Dublin.

Mercator's map of the British Isles on 8 sheets in 1564 includes Ireland, and the part relating to Ireland was copied by Ortelius as a separate map and first issued in his *Additamentum* of 1573 and was re-issued in the various editions of his Atlas up to 1595 (see Chapter V).

The Mercator-Ortelius map was superseded by the map compiled by Baptista Boazio (68). This was first printed about 1599–1600. Three copies are known, one in Dublin, one in the British Museum and one, in private hands, on silk. A copy of Boazio's map by Pieter van den Keere was published by Hondius in 1591. Another issue of Boazio's map appeared in 1599. This was engraved by Renolde Elstrack and sold by Mr. Sudbury. A facsimile was published by the B.M. in 1938. Boazio's representation was greatly superior to previous efforts, particularly in the south-west and north, and was adopted in the 1602 and subsequent editions of Ortelius. It was published by Vrints, and is slightly larger than the other maps in the atlas.

Mercator's separate map of Ireland first appeared in his atlas of 1595 (for later editions see Chapter V). It shows a quite different representation from that given in his map of 1564. Mercator's atlas likewise includes a 2-sheet map of Ireland, a separate map of the east part of Ulster and a map of Udrone. Mercator's map of Ireland was copied on a smaller scale in Camden's *Britannia* of 1607, engraved by W. Hole, and at a still later date on two sheets by Coronelli in 1695. The British Museum possesses a manuscript map of Ireland sent to the Earl of Salisbury by John Norden.

The 17th century opens with John Speed. His atlas, *Theatre of the Empire of Great Britain*, first appeared in 1611. The Fourth Book of this work deals with Ireland, and contains a general map and four maps of the four provinces (for later editions see Chapter VII), finely engraved by Jodocus Hondius. The general map has a large side panel of 6 compartments in which

67   RICHARD BLOME'S "IRELAND" ($15\frac{1}{4}'' \times 14\frac{1}{2}''$)

from his *Britannia*, 1673

68   BOAZIO'S *IRELANDE*, 1599

(32″ × 23½″) (B.M. Maps C2 cc 1)

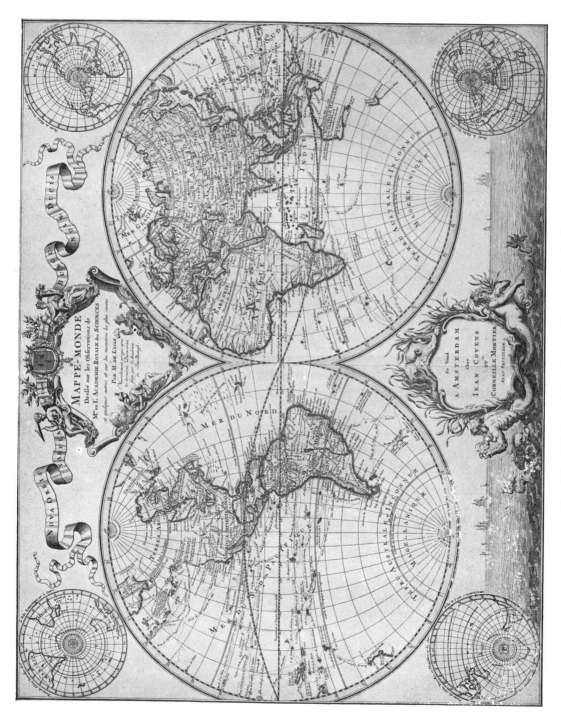

69 DE L'ISLE'S *MAPPE-MONDE* (25½″ × 19½″)

from Covens and Mortier's *Atlas*, c. 1740

are depicted 6 figures in national costume. A copy is known printed on satin, possibly the earliest use of this material for map printing. Munster has plans of Limerick and Cork, Leinster of Dublin, Connaught of Galway, and Ulster of Enis Kelling Fort. Speed's map is a great improvement on Mercator, particularly as regards the north coast, but the west coast is still defective. Upon whose survey Speed founded his map is not known, but Andrews suggests it may have been possibly on that of Robert Lythe of 1572. That Speed's representation was recognised as the best of its time is evident from its adoption by the Dutch school of map-makers, Blaeu, Jansson, Valck and Schenk, De Wit and Allard, the first four producing, not only his general map, but also those of the provinces. It was also utilised by J. Overton in his map of 1669 engraved by Hollar.

In 1646 Robert Dudley gave a chart of Ireland in his *Arcano del Mare*, which is based neither on Mercator nor Speed, though approximating more to the latter. It is weakest on the north and west coasts. The French school, starting with Sanson, gave yet another version substantially corresponding with Speed, but more elongated in form. A decorative map of Ireland was issued by Blome in 1673 (67).

A far better outline of the coasts of Ireland was given by the chart-makers in the marine atlases of the second half of the 17th century, first indicated by Waghenaer in 1588 and more fully developed by Jansson and Goos. Here for the first time a correct projection was given to the west coast of Connaught.

The most important event in the 17th century was, however, the production of Sir William Petty's *Hiberniae Delineatio*, the first atlas of Ireland. William Petty was born in Hampshire in 1623. He studied science and mathematics at Caen, Utrecht, Amsterdam and Leyden, returned to England and studied medicine at Oxford, and in 1652 was appointed Physician-General to the army in Ireland. His proposal to survey this kingdom was accepted by the Commonwealth Government against the opposition of the Surveyor-General Worsley. A capable organiser, Petty engaged 1,000 men to carry out this task, 40 clerks at headquarters, and a small army of surveyors and undermeasurers and a few "sagacious persons were employed to supervise and prevent scamping and frauds." The survey was completed and deposited in 1657. Petty's map of Ireland was engraved in Amsterdam in 1673, but the complete work did not appear till 1685.

The *Hibernatio Delineatio*, 1685, contains 36 maps, including a general map, 4 separate maps of the Provinces and 31 county maps. It was re-issued in 1690.

A reduction of Petty's atlas appeared in oblong 12mo in 1685 under the title *A Geographical Description of the Kingdom of Ireland* corrected and amended, and engraved by Fra. Lamb with 39 plates, published by Lamb, Morden and Berry. This was in its turn re-issued in 1689, 1695 and in 1728 by John Bowles, with roads added.

Petty's map quickly superseded all others, and was copied and utilised by succeeding map-makers both at home and abroad up to the issue of the ordnance survey. In Holland he was copied by De Wit (1680), Visscher (1690), Schenk (1690), Valck (1705), Allard (1710), Covens and Mortier (1730) and Van der Aa (1714); in Germany by Homann and Seutter about 1730; and in France by Jaillot, De l'Isle and Le Rouge. In England maps appeared by:

Morden and Overton (1680) and with additions (1683).
C. Browne (1691).
R. Morden 1695.
Philip Lea, 1699, engraved by Sutton Nicholls.
Moll (1700). Re-issued by T. and J. Bowles.
Moxon, 1704.
Pratt (1705), engraved by John Harris: re-issued by Grierson.
Price, Senex and Maxwell, 1718.
Moll, 1714; another edition by Grierson, 1733.

Kitchin, 1750.
Jefferys, 1759.
Bowles, 1760, 1779 and 1791.
Rocque, 1773.

Two atlases appeared in the 18th century: Moll's *Set of Twenty New and Correct Maps of Ireland*, in oblong 4to, 1728—this is extremely rare—and a small but charming *Hibernian Atlas*, by B. Scalé, was published by Sayer and Bennett in 1776 in quarto size, containing 37 maps. This was re-issued in 1778 by Sayer alone, and in 1798 by Laurie and Whittle. Kitchin and Faden both composed maps of Ireland in 1794 and 1798 respectively, and the *Roads of Ireland*, by Taylor and Skinner, in 1777 and 1783, deserves mention.

Four charts relating to Ireland appear in Captain Greenvile Collins' *Coasting Pilot*, viz. Carreckfergus Lough, showing Belfast, Carlingford, Dublin and Dublin Bay, and Kingsale Harbour.

Some finely engraved plans appeared in the latter half of the century:

Brooking (C.). Map of the City and Suburbs of Dublin, 3 sheets, 1728.
Rocque (John). Exact Survey of the City and Suburbs of Dublin, 4 sheets, 1756.
—— Reduction on 1 sheet, engraved by P. Halpin, Dublin, 1757.
—— Plan of City and Environs of Dublin, 1773.
—— Survey of the City and Suburbs of Cork, 1773.
—— Survey of the City of Kilkenny.
—— Survey of the City, Harbour, Bay and Environs of Dublin, 2 sheets, 1773.
—— Plan of the City and Suburbs of Dublin, 1787.
—— and Scale. City and Suburbs of Dublin, 2 sheets, 1773.
Richards and Scalé. Plan of the City and Suburbs of Waterford, 1764.

Large-scale maps of the counties began to appear about the same time, and continued into the 19th century:

Allen (W.). County Carlow, 1798.
Bald (W.). Mayo, 1830.
Bath (N.). County of Cork, 6 sheets, 1811.
Byron (S.). Plan of Dublin (1785).
Duncan (W.). Dublin, 1821.
Edgeworth (W.). Longford, 4 sheets, 1814.
—— and R. Griffith. Roscommon, 2 sheets, pub. Cross, 1825.
Gill (V.). Wexford, 4 sheets, Faden, 1816.
Larkin (W.). Meath, 1812–17.
—— Galway, 16 sheets, 1819.
—— Leitrim, 6 sheets, 1819.
—— Sligo, 6 sheets, 1819.
Lendrick (J.). County of Antrim, 4 sheets, 1782 and 1808.
McCrea (W.). Moneghan, 1795.
—— Donegal, 4 sheets, 1801.
—— and G. Knox. Tyrone, 4 sheets, 1813.
Nevil (A. R.). Wexford, 1798.
Nevil (J.). Wicklow, 1760.
Noble (J.) and J. Keenan. County of Kildare, 2 sheets, 1752.
Pelham (H.). Clare, 12 sheets, 1787.
Rocque. Map of the County of Dublin on 4 sheets (1760), re-issued Laurie and Whittle, 1802.
—— Map of the county of Armagh on 4 sheets, 1760).
Taylor (Lieut. A.). Map of the County of Kildare, 6 sheets, 1783.
Taylor and Skinner. Map of the County of Louth, 4 sheets, 1777.
Williamson (J.). Down, 4 sheets, 1810.
Wren (M.). County of Louth, 4 sheets, 1766.

In the 19th century maps became too numerous to mention individually, but the Ordnance Survey was commenced in 1829 and completed in 1858, consisting of 205 sheets and index, and the following atlases appeared:

Lewis's *Atlas* in 1837, in quarto, with 33 maps, re-issued in 1846.
Phillips' *Handy Atlas of the Counties of Ireland*, in 8vo, in 1881, 1885 and 1894.
Phillips' *Atlas and Geography of Ireland*, in 4to, in 1883.
Bartholomew's *Pocket Atlas of Ireland*, 1887.

AUTHORITIES

ANDREWS (M. C.). "The Map of Ireland, 1300–1700" (*Belfast Nat. Hist. and Philo. Soc.*, Belfast, 1923).

—— "Scotland in the Portolan Charts" (*Scottish Geogr. Mag.*, 1925).

—— "Boundary between Scotland and England in the Portolan Charts" (*Proc. Soc. Antiq. Scotland*, Vol. XII, Edinburgh, 1926).

—— "British Isles in the Nautical Charts of 14th and 15th Centuries" (*Geogr. Jnl.*, Vol. LXVIII, 1926).

BLAKE (M.). "Map of the County of Mayo in 1584 with Notes Thereon" (*Jnl. of Galway Arch. and Hist. Soc.*, Vol. V, No. III, Galway, 1907–8).

CARBERRY (Eugene). "The Development of Cork City (as shown by the maps of the City prior to the Ordnance Survey map of 1841–2)" (*Journal of Cork Hist. and Arch. Soc.*, Jan.-June, 1943).

CASH (C. G.). "Timothy Pont" (*Scottish Geogr. Mag.*, Vol. XVIII, 1901).

CHUBB (T.). *The Printed Maps in the Atlases of Great Britain and Ireland: a Bibliography, 1579–1870* (London, 1927).

COLBY (T.). *Ordnance Survey of the County of Londonderry: Memoir of City and N.W. Liberties* (Dublin, 1837).

COWAN (W.). *The Maps of Edinburgh, 1544–1929* (1932).

DUNLOP (R.). "16th-Century Maps of Ireland" (*English Hist. Review*, 1905).

FITZMAURICE (Lord Edmond). *Life of Sir William Petty, 1623–1687* (1895).

FORDHAM (Sir H. G.). *Notes on British and Irish Itineraries and Road Books* (Hertford, 1912).

—— *Road Books and Itineraries of Ireland, 1647–1850* (Dublin, Falconer, 1923).

GOUGH (Richard). *Anecdotes of British Topography* (1768, also 1780).

HAYES-McCOY (G. H.). *Ulster and other Irish Maps* (Dublin Stationery Office, 1964).

HEAWOOD (E.). "Mercator's Large Map of the British Isles and Lost Mercator Maps" (*Geogr. Jnl.*, Vol. LXII, London, 1923).

INGLIS (H. R. G.). "Early Maps of Scotland and their Authors" (*Scottish Geogr. Mag.*, Vol. XXXIV, Edinburgh, 1918).

LYNAM (E. W. O'F.). "Boazio's Map of Ireland" (*B.M. Quarterly*, Vol. II, No. 2, 1937).

MANLEY (G.). "The Plancius Map of England, Wales and Ireland, 1592" (*R.G.S.J.*, September, 1934).

MOWAT (J.). "Old Caithness maps and mapmakers" (*John O'Groat Journal*, Wick, 1938).

PETTY (Sir William). *History of the Survey of Ireland, commonly called the Down Survey*, by Doctor William Petty, 1655–6, edited by T. A. Larcomb (Dublin, 1841).

POWER (Rev. P.). "Old Map of Dungarren dated 1670" (*Jnl. Waterford and S.E. Ireland Arch. Soc.*, Vol. XIV, Waterford, 1911).

ROYAL SCOTTISH GEOGRAPHICAL SOCIETY. *The Early Maps of Scotland*. 3rd Edition, Revised 1973.

SHEARER (J. E.). *Old Maps and Map-Makers of Scotland* (Stirling, 1905).

SKELTON (R. A.). "Bishop Leslie's Maps of Scotland, 1578" (*Imago Mundi VII*, 1950).

TOOLEY (R. V.). "Maps in Italian Atlases of 16th Century" (*Imago Mundi III*, 1939).

WESTROPP (T. J.). "Early Italian Maps of Ireland from 1300 to 1600" (*Proc. of Royal Irish Academy*, Vol. XXX, Dublin, 1913).

CHAPTER X

Africa

AFRICA forms a half-way house in geography, Egypt and its northern coasts being known in remote antiquity, its southernmost tip not reached till 1487 by Bartholomew Diaz, and its centre practically unknown till the late 19th century.

In the early editions of Ptolemy, part of Africa is shown in the map of the world, and separate maps are devoted to its northern littoral and Egypt. There are maps of Africa in all editions of Ptolemy (see Chapter I), but its southernmost parts are not shown till the edition of 1508, and the first edition of Ptolemy to have a separate map for the southern area showing the Cape is the 1513 Strassburg edition. In the later editions of Ptolemy, maps of the whole of Africa appear among the tabula moderna.

The Cape, however, was depicted before that date, for example, on Behaim's globe (1492), the Laon Globe (1493) and the Lennox Globe of 1508. This is a remarkable achievement for the time, cartographical representation usually lagging far behind geographical discovery. Behaim, who constructed his globe in his native city of Nürnberg, had special facilities for his task, being employed in the Portuguese service. Even before this date a break with the Ptolemaic tradition of a landlocked Indian Ocean was shown in the works of Macrobius and Sacrobosco and in the portolan charts, for example, of Marino Sanudo 1306, the Laurentian portolan of 1351, an anonymous portolan of 1417, the Catalan map of 1450 and Fra Mauro's world map of 1457. All these show a theoretical connection between the South Atlantic and the Indian Ocean. The outline of Africa first began to approach its real shape in the manuscript maps of the first years of the 16th century; for example, in the world portolan of Juan de la Cosa and a finer map by Caneiro in 1502. The coastal nomenclature in Caneiro's map is extensive, 360 names being given. The padraos set up by the Portuguese to mark their claim to the country are shown round the coast, the interior is practically empty, though a loaded elephant led by an armed Moor is placed in the interior of South Africa. Cantino's chart appeared in the same year, and Pilestrina's portolan in 1503.

Of printed maps, the earliest so far known to exist showing the whole of Africa is a woodcut map on a newsletter *Den Rechtenweg*, published 1505. It was followed in 1506 by Contarini's world map, by Waldseemüller in 1507, and in 1508 by George Reisch in his world map to accompany the Ptolemy atlas of that date. Another early printed map of Africa is that which occupies practically the whole of the title-page of a work entitled *Itinerarium Portugalesium*, published in Milan in 1508. A year later, in 1509, a small map of Africa showing the Cape appeared in a work published in Strassburg—the *Globus Mundi*. Africa is also shown in the world map in Reisch's *Margarita Philosiphica* of 1513, in the *Carta Marina* of Waldseemüller of 1516, and in the world map of Grynæus of 1532. A popular map of Africa is that which

70 *Woodcut Map of Africa, Munster, 1540*

appears in the various editions of Sebastian Münster's *Cosmographia*. This map, printed from a woodblock, poor in its geographical content, is attractive to collectors, not only on account of its pictorial value, being adorned with small figures of a cyclop, elephant, native potentates, etc., but as being the earliest map of Africa that is generally available.

In the latter half of the 16th century maps devoted exclusively to Africa became more numerous, and the following appeared:

1554 (and 1563, 1588) Ramusio.		1575	Thevet.
1560	Gastaldi, engraved by J. and L. a Doeticum.	1578 (and 1593) De Jode.	
1562 (and 1564) Forlani.		1579 (and 1580) Duchetti.	
1563 (and 1566) Camocius.		1590	Pigafetta, engraved by Bonifacius.
1564	Gastaldi; a large map on 8 sheets (6 copies known).	1592	Plancius.
		1595	Mercator.
1564	Nicolo Nelli; on 3 sheets (Africa, Arabia, India, etc.).	1596	Linschoten.
		1597	Pigafetta, engraved by Wm. Rogers.
1565	Bertelli.	1597	Hulsius.
1570	Ortelius.	1600	Arnoldo di Arnoldi.
		1600	Quad. Bussemacher.

The maps of Africa by Ortelius and Mercator went into many editions. They are extremely decorative, with large title-pieces, usually to be found coloured, though uncoloured examples exist. One other important publication in the 16th century must be mentioned, the first African atlas, the *Geografia* of Livio Sanuto, published in Venice in 1588. This contains **12** maps.

The 17th century witnessed an enormous increase in publications devoted to maps, of which the following are a selection:

1602	Botero.	1664	Du Val.
1613	Ramusio.	1666	Chetwind.
1626–7	Speed.	1666	Hollar.
1630	Blaeu.	1666	Overton.
1631	Hondius.	1669	Sanson-Mariette.
1632	Jansson.	1670	De Wit.
1636	Visscher (N. J.).	1670	Ogilby.
1640	Bertius.	1671	Picart.
1646	Merian.	1680	Valk (6 sheets).
1646	Davity.	1680	Berry-Sanson.
1650	Blaeu.	1680	De Wit.
1650	Sanson.	1680	Overton-Lea.
1650	Jansson.	1680	Allard.
1652	Seile-Trevethen.	1682	Blome.
1655	Walton.	1685	Sanson-Jaillot.
1656	Sanson.	1689	Overton and Lea.
1656	Visscher.	1690	Valk.
1659	Hondius.	1690–6	Sanson-Jaillot.
1659	Cluver.	1700	Schenk.
1660	Nicolai (4 sheets).	1700	Moll.
1660	Danckerts.	1700	De l'Isle.
1663	Seile-Vaughan.	1700	Wells.

All these are highly decorative sheets, mostly issued in colour, and all from general atlases or works of travel or geography. Several of them, for example the Hondius, Blaeu (72), Jansson and Sanson, went into numerous editions (see Chapters V and VI). The most highly ornamented of this list are the maps by Blaeu, Speed, the 1636 Visscher, the Walton and the early De Wit, all of which have elaborate borders giving vignette plans of the principal cities

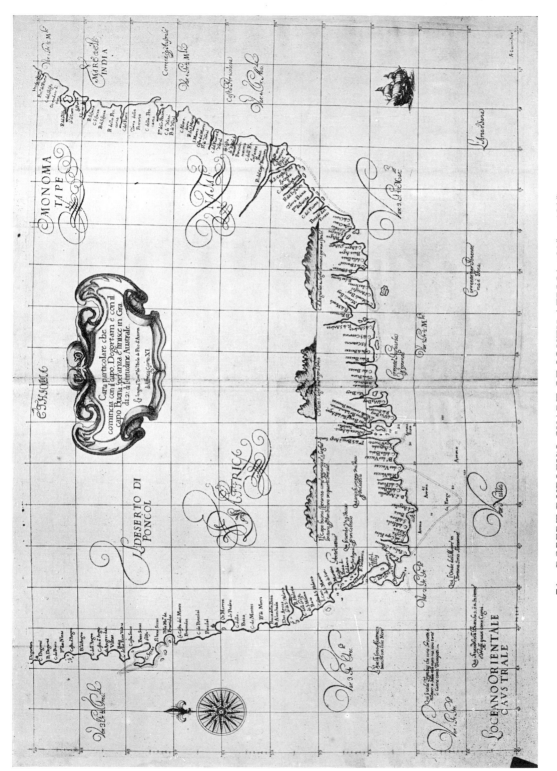

71   ROBERT DUDLEY'S CHART OF SOUTH AFRICA (29″ × 18½″)

from *Arcauo del Mare*, 1646

72   G. BLAEU'S MAP OF AFRICA

($22'' \times 16''$) from the *Grooten Atlas* (1648–65)

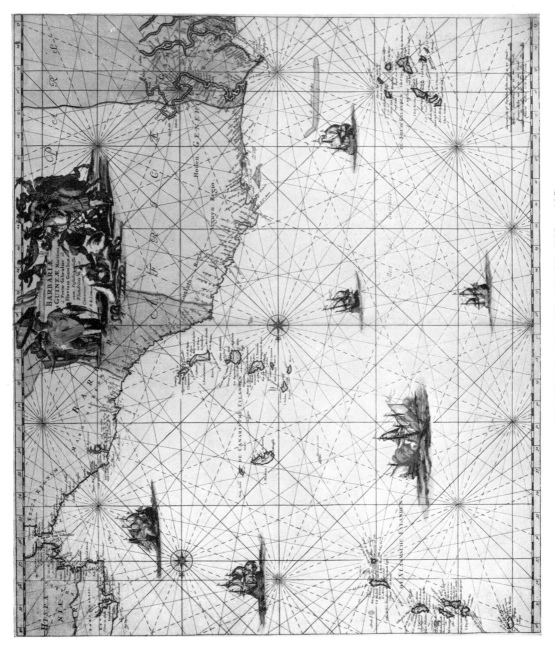

73 OTTENS'S BARBARY COAST (22″ × 19″)

from *Atlas van Zeevaert en Koophandel door de Geheele Weereldt*, 1745

and figures depicting the costume of the various inhabitants, as well as small figures, ships, animals, etc., on the body of the map. The map by Hondius is charmingly illustrated with ships, flying fish, etc., and all have large and decorative title-pieces. The map by Speed is the first English map of Africa.

The preceding are all land maps: there are, in addition, many marine charts from the hydrographical atlases, particularly those from the Dutch presses, the maps of Van Alphen, Blaeu, Colom, Doncker, Dudley (71), Goos, Jacobsz, Keulen, Ottens (73), Robijn, Seller, Voogt and Waghenaer.

The following maps of the 18th century may be cited:

1703	Mortier.	1749	Robert.
1704	Nolin.	1754	Nolin. Re-issued Denis, 1817.
1705	De Fer.	1756	Vaugondy.
1709	Zürner.	1760	Bowen.
1711	Price and Senex.	1760	Janvier.
1714	Van der Aa.	1763	Tirion.
1714	Moll.	1767	Robert.
1715	N. de Fer.	1769	Janvier.
1716	Thornton.	1772	Sayer after D'Anville.
1720	Senex.	1778	Vaugondy.
1721	Price-Willdey.	1780	Janvier.
1722	De l'Isle.	1780	Lotter.
1730	Homann.	1782	De la Rochette (74).
1730	Schenk.	1782	Janvier.
1730	Covens and Mortier.	1784	Zatta.
1735	Tirion.	1787	Harrison.
1737	Hase.	1787	Sayer.
1740	Seutter.	1790	Stackhouse.
1744	Bowen.	1794	Laurie and Whittle.
1745	De Lisle-Buache.	1794	Kitchin.
1745	Bakewell.	1795	De la Rochette-Faden.
1747	Le Rouge.	1795	Reilly.
1749	D'Anville.	1797	Güssefeld.
1749	Brouckner.	1800	Wilkinson.

The maps of the 18th century still maintain a decorative quality, though their ornamental side is usually confined to the title-piece and scale of miles. Many of the title-pieces have large figures depicting the natives and products of the continent; for example, Senex, Moll, Price, Homann, and Seutter, a favourite motif being a representation of a native riding a crocodile. The large map of Moll is of particular interest for the inset views it contains of the Cape of Good Hope, Cape Coast Castle, St. James Fort, etc. In the latter maps of this period it is possible to study the gradual extension of European penetration into the interior. As in the preceding century, there are marine charts of the coasts, though the emphasis is now on the French and English productions, a notable example being Part V of the English Pilot, the *West Coast of Africa from the Straits of Gibraltar to the Cape of Good Hope*, issued in 1701 and numerous subsequent editions up to 1780.

In the 19th century ornament, with one or two exceptions, for example the atlases of Tallis and Fullarton, was excluded, though many are remarkable for the excellence of their engraving. Their primary interest lies in their historical recording of the spread or penetration of European element in the interior, the gradual discovery and naming of physical features, the towns and trading posts established. A very fine map on four sheets was produced by Aaron Arrowsmith in 1802 (re-issued in 1811), and a four-sheet map by Purdy in 1809. Other excellent maps on a smaller scale were issued by Cary in 1805, 1811 and 1828, and in France by Brué (1800), Lapie (1803) and De l'Isle-Dezauche in 1805; a German map by Güssefeld in 1804, and an American map by Carey in 1817; also Wyld's re-issue of Faden's map in 1835 and 1850.

Separate maps of the various parts of Africa were issued from the 16th century onwards, several charming maps being executed of the islands of the Azores, Cape Verde, Canary, St. Helena and Madagascar, by Ortelius, Blaeu, Jansson, Sanson, Ogilby and others, as well as the marine map-makers Dudley, Goos, Mount and Page, etc.

EAST AFRICA

The east coast was the last portion of the continent to be developed by Europeans, and in consequence has the fewest maps devoted to its representation. One of the main centres of interest in early times was Abyssinia, thought to be the land of Prester John, and so a special object of interest to Christendom, and many of the early maps from Ortelius, Blaeu and Ogilby downwards give representations of this area, though stretching far beyond the actual confines of that kingdom. The map of Ortelius, "Presbiteri Johannis sive Abissinorum Imperii descriptio," largely based on Jesuit relations, embraces the whole of East and Central Africa, from the Egyptian border in the north to Mozambique in the south.

A very brief list of maps showing parts of the east coast is as follows:

1572	Ortelius.	1728	Mount and Page.
1596	Linschoten and Langeren.	1747	Bowen.
1640	Blaeu.	1767	Bellin.
1650	Jansson.	1782	Bonne.
1655	Sanson.	1801	Reinecke.
1670	Ogilby.	1814	Pinkerton.
1680	De Wit.	1814	Salt.
1683	Valk and Schenk.	1851	Tallis.
1700	Mortier.	1860	Fullarton.
1714	Moll.	1874	Livingstone.
1715	Renard.	1878	Stanley.

NORTH AFRICA

Egypt and North Africa are rich in cartographic material. In all the editions of Ptolemy, in the atlases of the Dutch, French and English schools, maps of North Africa are to be found. It is likewise rich in MS. material, all the portolan maps starting with the Mediterranean as their base. The atlases of Ortelius, Mercator, Blaeu, Jansson, De Wit, Visscher, Sanson, d'Anville, De l'Isle, Seller and others all contain decorative coloured maps, either of the whole of North Africa or its various kingdoms, Fez, Morocco, Algiers, etc.; and prior to the publication of the first standard atlas of Ortelius, many maps of the northern coast of Africa were included in the Italian collections; for example, plans of Algiers in 1541 by Salamanca; 1565 Bertelli, and 1579 Duchetti; of Madagascar in 1567 by Gastaldi, and of Tunis in 1535 by Agostino Veneziano; 1560 Zalterius, and 1602 Orlandi.

A brief list is as follows:

1540	Münster.	1702	Seller and Price.
1572	Ortelius.	1720	De l'Isle
1595	Mercator.	1730	Homann.
1640	Blaeu.	1735	Seutter.
1649	Jansson.	1771	Bonne.
1680	De Wit.	1802	Reinecke.

WEST AFRICA

West Africa has likewise an extensive series of maps devoted exclusively to its area, from the manuscript charts of the first Portuguese explorers down to present times, from the various editions of Ortelius, Mercator, Blaeu and Jansson, Blome, Ogilby and Seller, and in the numerous sea atlases of the 17th century.

The following brief list may be cited:

1596	Langeren and Linschoten.	1730	Homann.
1606	Mercator and Hondius.	1743	D'Anville.
1640	Blaeu.	1750	Bellin.
1650	Jansson.	1760	Bowen.
1655	Sanson.	1768	Jefferys.
1670	Ogilby.	1771	Bonne.
1680	Robijn.	1781	Bonne.
1693	Blome.	1793	Ehrmann.
1700	Schenk and Valck.	1802	Norie.
1704	De Fer.	1830	O'Beirne and Major Laing; Wyld.
1726	De l'Isle.	1831	Weiland.
1730	Ottens.	1840	Tallis.

SOUTH AFRICA

Maps of South Africa are somewhat naturally the most popular and interesting of all the regional maps of the continent. The earliest in date give coastal detail only, such as that in the 1513 edition of Ptolemy, the first separate map devoted to South Africa. This is extremely rare and unlikely to come the way of collectors. Other maps, or rather charts, such as the Langeren, Goos and Mount and Page, show the whole of the coast-line from Guinea to the Cape. The most decorative examples commence with the map of Blaeu, which has a handsome title-piece with Hottentot supporters, and similar maps of a smaller size were issued by Merian and Ogilby. Some, such as the Thornton, the Seller and the Norie, are distinguished by inset views of Table Bay and the Cape.

A brief list is as follows:

1598	Linschoten.	1780	Lotter.
1640	Blaeu.	1785	Sparrman.
1646	Merian.	1788	Bonne.
1646	Dudley.	1790	Ruyter; Sayer (inset view of Cape).
1652	Jansson.	1795	De La Rochette (74).
1655	Sanson.	1797	Ehrmann.
1669	Goos.	1797	Rennell.
1670	Ogilby.	1802	Reinecke.
1675	Seller.	1807	Weimar Geogr. Inst.
1680	De Wit.	1810	Gottholdt.
1683	Robijn.	1815	Thomson.
1700	Keulen.	1818	Walker-Horsburgh.
1702	Mount and Page.	1827	Owen.
1715	Renard.	1831	Norie.
1720	De Lisle.	1834	Owen.
1728	Mount and Page.	1840	Wyld.
1730	Ottens.	1842	Arrowsmith.
1740	Moll and Bowles.	1850	Tallis.
1745	De Lisle; Buache.	1850	Wyld.
1754	Bellin.	1853	Cox.
1757	Bellin.	1857	Hall.
1760	Bowen.	1860	Fullarton.
1763	Tirion.	1889	Juta.

AUTHORITIES

ALMAGIÀ (R.). *Il contributo di Venezia alla conoscenzi dell' Africa* (Venezia, 1938).

BERCHET (Cav. G.). "Lettera sulla cognizioni chi i veneziani avevano dell'Abissinia" (*Boll. della Soc. Geogr. italiana*, Firenze, 1869).

CORTESÃO (A.). *Cartografia e Cartografos portugueses dos seculos XV e XVI* (Lisbon, 1935).

—— and E. HENRY THOMAS. *Carta das Novas que vieram a el rei nosso senhor do descobrimento do Preste João, Lisboa, 1521* (Lisboa, 1938).

DENUCÉ (J. B. F.). "Africa in de XVI Eeuw en de Handel van Antwerpen" (*Dok. voor d. Geschiedenis van den Handel*, Antwerpen, 1927).

—— *L'Afrique au XVI Siècle et le Commerce Anversois*, Anvers, 1937.

—— "Les sources de la carte murale d'Afrique de Blaeu de 1644" (*Cong. Géogr. Int. Comptes rendus*, Leyden, 1938).

FISCHER (J.). "Abessinien auf dem Globus des Martin Behaim von 1492" (*Peterm. Mitt.*, 1940).

HOPPER (Stephanie). *Miniature Maps of Africa*. Map Collectors' Circle, 1975.

JOHANNESBURG PUBLIC LIBRARY. Descriptive Catalogue (1952).

KOEMAN (C.). *Tab. Geograph. quibus Colonia Bonae Spei antiqua depingitur* (Amsterdam, 1952).

LANE-POOLE (E. H.). *The Discovery of Africa . . . as reflected in the maps in the collection of the Rhodes Livingstone Museum* (Livingstone, 1950).

LEFEVRE (R.). "L'Abissinia nella cartografia medievale" (*Le Vie del mondo*, 1940).

—— "L'Africa orientale nella cosmografia patristica e nella cartografia genovese del Trecento" (*Riv. d. Col. Roma*, 1939).

LYNAM (E. W. O'F.). "The Discovery of Africa (Mon. Carto. Africae et Aegypti)" (*B.M. Quarterly*, 1927).

MAJOR (R. H.). "On the map of Africa published in Pigafetta's Kingdom of Congo, 1591" (*R.G.S. Proc.*, 1867).

MENDELSSOHN (S.). *South African Bibliography . . . Cartography* (pp. 1095–1113) (2 vols., London, Holland Press, 1957).

OGUNSHEYE, (F. A.). "Maps of Africa, a Bibliographical Survey", (Ibadan, *Nigerian Geographical Journal*, 1964).

PHILLIPS (P. Lee). *List of Geographical Atlases in the Library of Congress*, 4 vols., 1909–20.

RANDLES (W. G. L.). *S. E. Africa and Empire of Monomotapa* (Lisbon, 1958).

RUGE (S.). *Topographische Studien zu den portugesischen Entdeckungen an den Küsten Afrikas* (Leipzig, 1903).

SANTAREM (Vicomte de). *Recherches sur la priorité de la Découverte des Pays situés sur la côte occidentale d'Afrique*, 8vo, Paris, 1842.

—— *Demonstracão dos Direitos que tem Portugal sobre . . . Molembo Cabinda e Ambriz*, 1855.

SCHRIRE (D.). *Cape of Good Hope 1782–1842: from De la Rochette to Arrowsmith* (London, Map Collectors' Circle, 1965).

TEIXEIRA DA MOTA (A.). *Cartograf. Antiga da Africa central 1500–1860* (Lourenço Marques, 1964).

TOOLEY (R. V.). *Printed Maps of the Continent of Africa and Regional Maps South of the Tropic of Cancer* (Parts 1 & 2, London, Map Collectors' Circle, 1966).

—— *Collectors' Guide to the Printed Maps of Africa* (London, Carta Press, 1969).

—— *Printed Maps of Southern Africa and its Parts* (London, Map Collectors' Circle, 1970).

—— *A Sequence of Maps of Africa* (with pictorial borders). Map Collectors' Circle, 1972.

YOUSSOUF KAMEL. *Monumenta Cartographica Africae et Aegypti* (13 vols., Cairo, 1926).

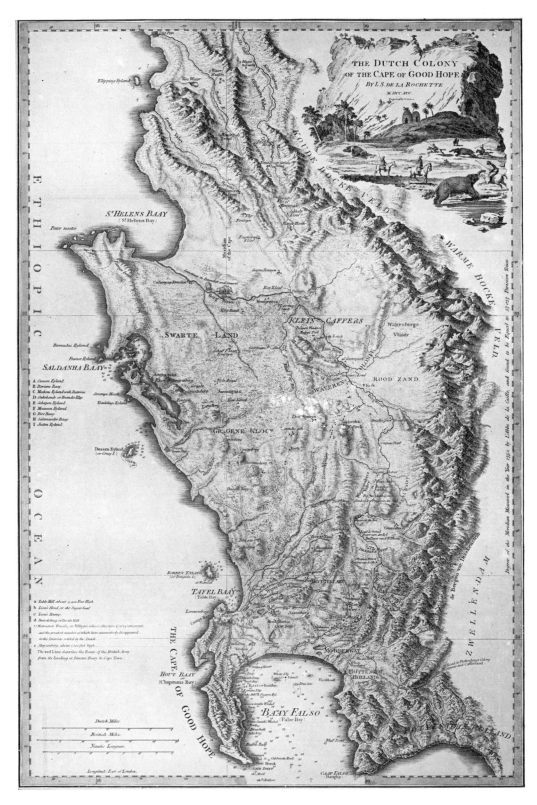

74   L. S. DE LA ROCHETTE'S MAP OF THE CAPE OF GOOD HOPE (14″ × 19½″)
from Faden's *General Atlas*, 1795

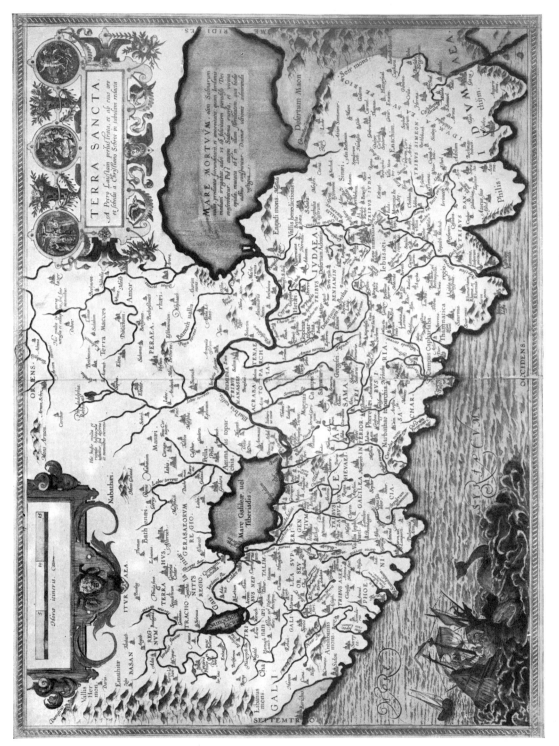

75   ABRAHAM ORTELIUS'S "PALESTINE" (20" × 14¼")

from Theatrum Orbis Terrarum, 1570

CHAPTER XI

Asia

ASIA, as part of the ancient or classical world, has no less than 12 maps devoted to its
delineation in Ptolemy's atlas (for editions see Chapter I) covering the whole of Asia,
with separate maps of the various kingdoms and empires except the farthest east. In
the later editions of Ptolemy in the tabulæ modernæ, new maps based on contemporary knowledge were added. General maps of Asia as well as separate maps of the various states may be
found in the general atlases of the Italians, Dutch, French and English schools (see respective
chapters). An outstanding separate map of Asia was that by Gastaldi on 3 copper-plate sheets
1559–61. This went into several editions, and was copied in whole and in part by other
cartographers, Ramusio and de Jode among others.

PALESTINE

Palestine differs from other countries. As the centre of Christian aspirations, it naturally
excited more interest at an early date than any other land, and consequently more maps have
been devoted to its representation than that of any other land in the Orient. It was a fairly
common practice to insert maps of Palestine in early Bibles, as for example that in a Dutch
Bible in the British Museum dated 1538. Pictorial representation of the Holy Land goes back
of course to a much earlier date, for example, the Madaba mosaic of A.D. 550. The introduction
of printing greatly increased the volume of maps: a map of Palestine appears in the *Rudi-
mentum Noviciorum* of 1475. In 1532 Jacobus Ziegler produced his *Quae intus Continentur Syria
&c.* This work, published in Strassburg, includes 7 maps of Palestine and 1 of Scandinavia. It
was re-issued in 1536. A map on a larger scale and of greater importance was that published
by Mercator in 1537. Printed on 6 sheets, only 1 example is known to have survived to the
present day. It is preserved in the *Biblioteca Civica* in Perugia.

In the collected Italian atlases of the 16th century, Palestine was well represented, maps
by the following cartographers appeared: Salamanca 1548, Della Gatta 1557, Mario Kartaro
1563, Forlani 1566, Zalteri 1569, and Jenichen 1570. All these are rare, though known from
several examples.

In France, Guillaume Postel produced a map on 2 sheets in 1562, and in Holland a large
map on 9 sheets by Laicksteen-Sgrooten appeared. This was engraved by the brothers J. and
L. van Docticum and published by H. Cock.

The Dutch atlases frequently give more than one map of Palestine; for example, Ortelius
gives three versions of the Holy Land, one by Peter Laicksteen (75), and two by Tileman
Stella.

In 1593 another work of note appeared, the *Theatrum Terrae Sanctae* of C. Adrichom, illustrated with 12 maps, including general map, plan of Jerusalem, and maps of the locations of the various tribes. The map of Adrichom was extensively copied by later geographers.

In 1590 De Jode issued a map and in 1610 Hondius. The first printed English map was that by John Speed in 1617. This map of Speed's did not appear in his Atlas (*The Prospect of the World* was first issued in 1627 and did not contain a map of Palestine), but with the "Genealogies" recorded in the *Sacred Scriptures* by J. S(peed) 1617. The first issue was engraved by Renolde Elstrack, with printed text on the reverse headed, "An alphabetical table of Canaan." Speed's map was re-issued in the 1676 edition of his atlas with Elstrack's name removed and dated 1651, though on the title of the atlas it is stated that it is its first appearance. Other early maps are those by Kaerius (1620), Picart (1637), J v. Doet (1644) and Mariette (1646).

From the 17th century onwards various works were issued dealing with the ancient world; for example, by Jansson, Cluverius, Cellarius and Horn. Such works were popular and went into many editions. They all include both general and particular maps of Palestine. Perhaps the one with the most quaint and decorative series of maps is that in Thomas Fuller's *Pisgah Sight of Palestine*, issued in 1650.

India

In the earlier editions of Ptolemy the Indian peninsula is shown greatly foreshortened, and Ceylon is sometimes to the right, sometimes to the left in early maps. The shape of India was corrected in Reisch's world map of 1508 and in the tabula moderna of the later editions of Ptolemy. A map of the Indian Empire by Bertelli appeared in Venice in 1565. In the atlas of Ortelius, India is not represented as a single entity, but grouped with the East Indies. It is decorated with the arms of Portugal, ships and sea monsters. Mercator's map likewise covers a wide area, and India is shown in an even more attenuated form.

The first English map of the Mogol Territories was published by T. Sterne, globemaker, in 1619. It was drawn by William Baffin on information supplied by Thomas Roe, and was engraved by Renold Elstrack. This map was copied by Purchas in 1625, and in Paris in 1663. Other maps appeared in Holland by Hondius, Jansson, Blaeu, Visscher, Van Keulen and De Wit. In France by Sanson, Jaillot, De Lisle, D'Anville and Bellin; in Germany by Seutter and Homann, and in England by Moll, Bowen and others.

The greatest of the early cartographers of India was James Rennell, born 1742, died 1830. He joined the navy, but was lent to the East India Company and appointed Surveyor-General of Bengal in 1764 at the age of twenty-one years. He retired in 1777 after thirteen years' surveying. The *Bengal Atlas* came out in 1779. It contains 21 maps and plans. His second great work was the construction of the first approximately correct map of the whole of India, first issued in 1782. The merit of Rennell's work was quicky appreciated, and remained the standard for many years. The following are some of his maps:

> Map of Hindostan, 2 sheets, 1782. Re-issued Laurie and Whittle, 1794.
> Map of Bengal, Behar, Oude, Allahabad and part of Agra and Delhi, Dury, 1776. Re-issued Faden 1786, Laurie and Whittle 1794, Wyld 1824.
> An Actual Survey of the Provinces of Bengal, Behar, etc., Dury 1776, Laurie and Whittle 1794.
> Map of the Provinces of Delhi, Agrah, Oude and Allahabad, Laurie and Whittle, 1794.

Other surveyors after Rennell were Colonel Kelly, Captains Pringle, Allen, Wersebe, Lieutenant Macartney, Wm. Bret, H. Watson and Kinnear.

An important publisher at the beginning of the 19th century was Aaron Arrowsmith. He issued a six-sheet map of India in 1804 and an *Atlas of South India* on 18 sheets in 1822. Another of the later Indian atlases was that published by J. Horsbrugh in 1829. Colonel

Thuillier (later General Sir H. L.), surveyor-general, produced a map of Bengal and North-West India in 1860. This was re-issued in 1870, 1889 and 1890.

CHINA

China, with her ancient civilisation, has a long map history. Centuries ago her geographers were constructing maps, not according to Western scientific standards—for it is only in comparatively recent years that her scholars have admitted the sphericity of the globe—but nevertheless following set rules; and in a methodical manner the whole of China and its bordering lands were mapped, the former in some detail.

The earliest-recorded map in Chinese literature is that mentioned by the historian Ssu Ma Chien, who relates that a map was presented by the Prince of the State of Yen to the Prince of the State of Ch'in in 227 B.C. He also states that when the self-styled First Emperor was buried under Mount Li, his body reposed in a specially prepared chamber contrived like a world in miniature, "with the aid of quicksilver, rivers were made, the Yang-tsze the Hoang-ho, and the great ocean, the metal being poured from one into the other by machinery. On the roof were delineated the constellations of the sky, on the floor the geographical divisions of the earth" (Giles's *History of Chinese Literature*).

Many maps are recorded of the Han Dynasty (200 B.C.–A.D. 200) executed on wood and silk, and after the invention of paper by Ts'ai Lun, A.D. 105, this new material was also utilised.

The first great Chinese cartographer was P'ei Hsiu (A.D. 224–71), called by Sir Alexander Hosie "the father of Chinese cartography." Like Ptolemy in the Occident, he gathered together and studied all available existing information, reduced it to order, and propounded rules for the making of maps. He constructed a map of China on a scale of 500 li to an inch (a li is roughly one-third of a mile). This map (on 18 sheets) was deposited by the Emperor Wu Ti among the secret archives. It was accompanied by a written account of the "Six Principles of Cartography": 1. Rectilinear divisions (divisions of the map into equal squares). 2. Orientation. 3. Mileage. 4. Altitudes (the high and the low). 5. Right and oblique angles. 6. The curving and the straight.

During the T'ang dynasty an even larger map was prepared, 33 by 30 feet, a scale of 100 li to the inch. This was commenced in 785 by Chia Tan and completed in 801. Shen Kuo (1030–93), under the Sung dynasty, was the first to produce a map in relief. It was made in wax.

Of these and other early monuments, none is known to survive. The oldest-known extant Chinese maps are engraved on two stone tablets preserved in the "Forest of Tablets" at Hsianfu. They were engraved in 1137. One, called the "Footsteps of Yu," is supposed to have been engraved six months prior to the other.

At the commencement of the 14th century Choo Sze Pun compiled the Kwang-yu-too, and the maps from this work were used later by the Jesuit Father Martini.

The first European map of the world circulated in China was composed by the Jesuit Father Matteo Ricci, the Apostle of China. It was printed at Shao-King in 1584. A second map was issued at Nanking in 1599, and a third in Pekin in 1602. Of this third map four editions are known, one spurious. The Shao-King map of 1584 was merely a European map in Chinese guise, its object being to introduce Western science to the Chinese. It was based mainly on the work of Ortelius. The Pekin map, on the contrary, was specially constructed by Ricci with the prime meridian 170° East to bring China towards the centre of the map, a measure more likely to win approval in Chinese circles. This map was printed from wood-blocks on thin Chinese paper, and its measurement was roughly 12 by 6 feet. Father Ricci's map, as regards China and Japan, was a distinct improvement on contemporary European maps, Korea being correctly shown as a peninsula, and a twist being given to the islands of Japan instead of the usual perpendicular line. The map was covered with extensive "Legends" or notes, as, for example: "*Japan is a large island . . . brute force is the controlling factor, and although there is a sovereign over the whole country, the real power is in the hands of his vassals . . . very little value is attached as a rule to precious stones: gold and silver and old porcelain are esteemed much more highly.*"

The success attending Ricci's map and the favour it found in high official quarters led the Jesuits to produce another world map in China. This was undertaken by Father Ferdinand Verbiest, a Flemish Jesuit, about 1670. He produced a large map of the world in two hemispheres printed on six rolls, with two additional rolls of letterpress, for the Emperor Kang Hsih (1661–83).

It is now necessary to turn back a considerable period of time to discuss the European conception of China. China was known in classic times, not in any exact manner, but as a large mysterious land in the Far East. It was included by Ptolemy in his geography A.D. 150, and in the early editions of his work printed towards the end of the 16th century China is shown as Sinarum Regio and Cathay.

The first information of geographical importance on China obtained by a European was due to the labours of Marco Polo. He travelled extensively in the Far East in 1271–95, and enjoyed special facilities for observation through his favour with, and employment by, the Emperor Kublai Khan (Chi Tsou). Marco Polo left no map, but from the written account of his travels, later geographers were able to make the first faint beginnings in the reformation of the maps of these regions.

The map of China began to take definite, if inexact, shape in the 16th century. This was due primarily to the voyages of Portuguese traders and Jesuit missionaries. An important map of China, the first to appear in a European atlas, was issued by Ortelius in the Additamentum III to his *Theatrum Orbis Terrarum*, published in 1584. Compiled by Ludovico Georgio, a Portuguese Jesuit, this map remained the standard type for the interior of China for over sixty years (78). It was used by Linschoten, de Jode, Mercator and Hondius. In this map the limits of China were given as the Great Wall in the north and Cauchin China in the south. A distinctive feature was the placing of five immense lakes in the interior, one of these forming the western boundary of the country. Relatively correct in the south, the coast showing a northeast trend, the map becomes less trustworthy and decidedly inaccurate in the north and east. From Shantung to Che Kiang the coast is mapped in a straight line from north to south. Korea is not shown. This flatness of the coast of China was common to almost all the maps of China during this period, among others those of de Jode and Mercator. Linschoten in 1596 was one of the first to indicate the curve of the China coast, for in his map of the East Indies he shows the land stretching north and west above Nanking. On the other hand, Linschoten inserted Korea as a large fat island off the coast of China. Mercator likewise inserted Korea, but in this case as a long narrow island, and his east coast runs in an almost perpendicular line from Shantung to Fokien. Hondius, in his map of Asia 1631, inserts Korea as a peninsula for the first time, but still retains the five large lakes in the interior in common with the maps of Ortelius, Linschoten and Mercator.

The second landmark in the history of the European mapping of China was the appearance of the *Atlas Sinensis* in 1655. This was compiled by Father Martino Martini, an Italian Jesuit, who died at Hangtcheou in 1661. Based on Chinese sources, it was far in advance of any previous European work. For the period it was remarkably accurate, being the first to show a more correct eastern coast-line with the Shantung promontory. Published in Amsterdam in 1655, it was incorporated at that date, and in the later editions of the "Great" atlas issued by Blaeu. It consisted, besides text, of a general map of China, 15 maps of individual Chinese provinces, and a general map of Japan. As was usual with Blaeu's publications, it was offered for sale both plain and coloured. The *Atlas Sinensis*, apart from the technical excellence of its production, is important as being the first European atlas of China. It remained the

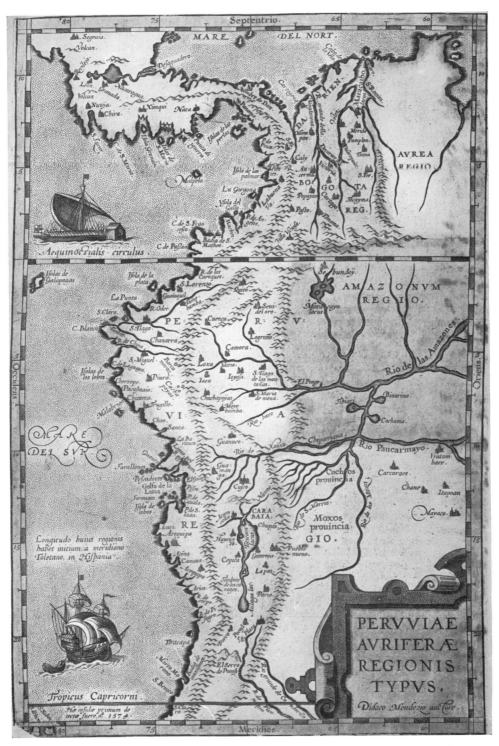

76   A. ORTELIUS'S PERU (19″ × 15″)

from *Theatrum Orbis Terrarum*, 1574

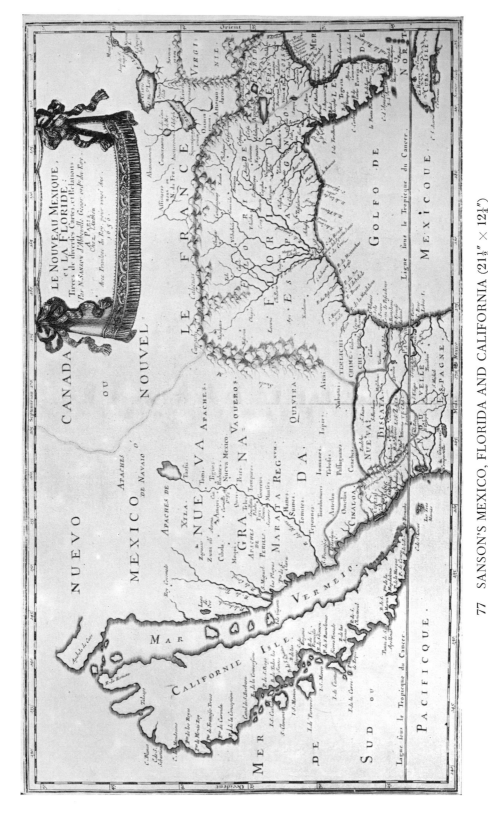

77   SANSON'S MEXICO, FLORIDA AND CALIFORNIA ($21\frac{1}{2}'' \times 12\frac{1}{4}''$)

from the *Cartes Générales* 1654

standard geographical work on that country till the publication in 1737 of D'Anville's *Atlas de la Chine*. This was also based on original surveys by the Jesuits.

As was usual in most early European maps (16th–18th centuries), all the preceding maps were gaily decorated, maps at that time being valued, not only for purely geographical reasons, but also as imparting even if in slight degree pictorial information on the habits and customs of the countries delineated. For example, the map of Ortelius is decorated with elephants and deer, Tartar tents, ships and sea monsters, and a number of wind carriages or boats for sailing on land. De Jode adorned his map with ships, and in his border vignettes shows Chinese cormorant fishing, idols and wind carriages. Mercator brought his embellishments up to date, inserting an illustration of a native junk as well as European ships, a picture of a crucifixion of a Christian in Japan, and a drawing of a Chinese wind carriage (this last copied from de Jode, only reversed).

With the publication of the *Atlas Sinensis*, a new type of decoration arose. The title, instead of being enclosed within interlacing strapwork ornament, was now surrounded by large figures depicting the costume of the area shown intermingled with various products of the same. Embellishments in the map itself were rendered far less prominent or discarded altogether, decoration being confined to the large plaque or cartouche containing the title of the map. Sometimes an exception was made for the scale of miles, this also being treated pictorially, as in Van Loon's map of China. An important work in the 18th century was Du Halde's *Description de la Chine*, 4 vols., folio, 1735, illustrated with maps and views. Two years later, namely in 1737, appeared the *Nouvel Atlas de la Chine, de la Tartarie Chinoise, et du Thibet*, by the celebrated geographer D'Anville, based on the Jesuit surveys of 1708–16, made for Kang-hsi. This contained 42 maps, and remained the principal cartographical authority on China during the rest of the 18th century. It is still possible to purchase examples of the works of Martini, Du Halde and D'Anville, and collectors may supplement these with maps from various atlases from Ptolemy down to modern times.

JAPAN

The first-known mention of map-making in Japan is made in the Nihongi, in an imperial edict of A.D. 646, ordering the execution of local surveys for the use of central authority. In the succeeding century numerous estate and administrative maps were compiled.

Gyogi Bosatsu (670–749) is, according to tradition, the most venerable figure in Japanese cartography. He was a Buddhist priest of Korean descent who migrated to Japan in his youth. He is said to have travelled extensively in Japan and helped the country people in the construction of roads, canals and bridges. He is credited with the construction of a map of the country. The earliest-known extant map of Japan is, however, one in the possession of the Temple of Ninna ji, near Kyoto. This map is presumed to be of the Gyogi type, has the south at the top of the map, and dates from 1305. The Gyogi type retained its influence for eight centuries, and had an effect on Chinese and European maps. Towards the end of the 16th century Japanese scholars came into contact with European map-making, and in the first part of the 17th century this was reflected, by Japanese artists, in the execution of world maps, several of which, drawn on screens during this period, have survived. The isolation policy followed by the Government later led to a decline in geographical knowledge, and a forward move was not made till Ishikawa Toshiyuki (Ryusen), who flourished 1688–1713, and compiled general maps and town plans. Ryusen's maps, beautifully executed—for he is said to have been a pupil of Moronobu—were nevertheless inferior cartographically to some of the older examples.

The next name of importance is that of Sekisui, who in 1779 compiled a general map of Japan which considerably influenced his successors. Sekisui was the first Japanese to employ meridians as well as scale in his map.

In the early part of the 19th century attention was turned to the north, and Mogami Tokunai and Takahasi Sakusaimon made maps of Jezo. The most notable of the later map-makers, prior to modern times, was, however, Ino Tadutaka (Chukei), who between 1800 and 1816 made an excellent detailed survey of the whole country.

Japan was not known to Europe till a far later date than China, and European maps of the country did not appear till the end of the 16th century. Ortelius, in his map of the East Indies, 1570, drew Japan as a fat little island something like a kite, with a long tail of minute islands at its head and tail. As if not entirely satisfied with this configuration, he gave an entirely different figure to Japan in his map of Asia published in the same year and the same atlas. In this case Japan was elongated and lost its upper tail. Mercator's map of Asia followed the same model for Japan. De Jode, in his map of Asia, gave yet another variation for Japan. He mapped it roughly in the shape of a thick crescent with two horns pointing due south. Linschoten, 1595–6, used the same conception.

In 1595 Ortelius published his first separate map of Japan. This was compiled by Father Ludovico Texeira, and is important to the collector of Japanese maps as being the standard European map of Japan for many years. It was based on the Gyogi type and its title is "Japoniae Insula Descriptio Ludovico Teisera auctore." It shows the island of Bungo running north and south and the larger island Japonia running due east and west. Korea is shown as a narrow island to the east.

The Texeira-Ortelius map was re-engraved by Hondius. He omitted the author's name and contracted the title to the single word "Japonia." It was practically an identical copy of Ortelius as regards the map itself, two names—"Fongo" and "Allias"—being added, one omitted and two changed. The most noticeable alteration was the addition of a long inscription within the island of Korea, and the changing of the decorations by the substitution of a native ship for one of the European vessels depicted by Ortelius. Mercator's variant of the Texeira map was again re-issued by Jansson about 1650, without any change save in the decorations and the title, which was then amplified to "Japoniae nova descriptio."

About the same time the Jesuit Father Briet compiled a new map of Japan which showed a distinct advance on the Texeira map. He altered some, and added more, place-names, marked internal boundaries, and considerably changed the contour of the coast-line.

Just at the turn of the century, Schenk and Valck printed a map of Japan after Jansson, and this map was one of the earliest to show land to the north of the main island of Japan. This accretion was called "Landt van Eso," and a considerable gap separated it from the mainland. Korea on this map was still marked as an island.

In 1750 the Sieur Robert published a map of the Japanese empire, and this at last began to show the more correct north-eastward sweep of the islands, with Korea correctly given as a peninsula.

AUTHORITIES

AHLENIUS (K.). "En Kinesick Värlskarta fran 17 århund" (*krifta af Kongl. Humanistiska Vetenskaps-Samfundet i Upsala*, Vol. VIII, 1902–4).

BADDELEY (J. F.). "Father Matteo Ricci's Chinese World Maps, 1584–1608" (*R.G.S.J.*, Vol. L, 1917).

BAKER (J. N. L.). "Some original maps of the East India Co." (*Konin. Nederl. Aardrijkskundig Genootschap 2 Serie deel LIII*, 1936).

BEANS (G. H.). *A List of Maps of the Tokugawa Era* (Tall Trees Library, Jenkintown, 1963).

CHAVANNES (E.). *Bulletin de l'Ecole Française de l'Extréme Orient*, Avril-Juin, 1903.

CRESSEY (G. B.). "Evolution of Chinese Cartography" (*Geogr. Review*, 1934).

CRONE (G. R.). *Seventeenth Century Dutch charts of the East Indies* (Geogr. Jnl., 1943).

DAHLGREN (E. W.). "Les Débuts de la cartographie du Japon" (*Arch. d'Etudes Orientales, Upsala*, 1911).

DENUCE (J.). *Magellan. La Question des Moluques et la Première Circumnavigation du Globe* (Brussels, 1911).

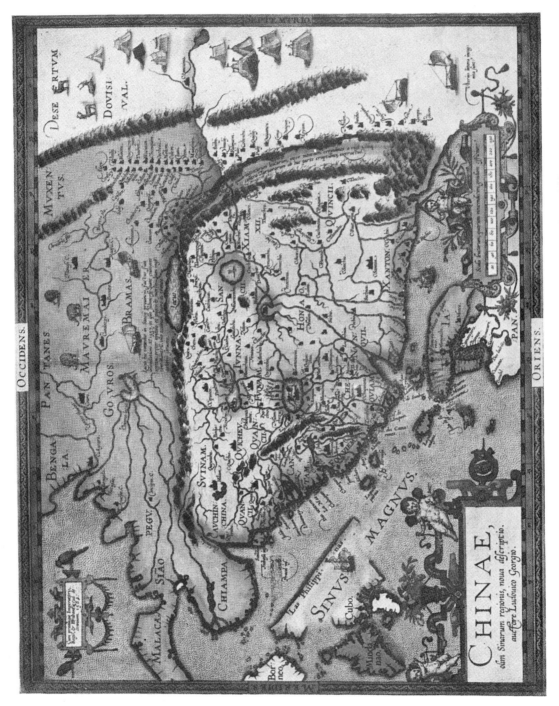

78 LUDOVICO GEORGIO'S MAP OF CHINA (18½″ × 14½″)

from Ortelius's *Theatrum*, 1584

GERINI (Col. G. E.). *Researches on Ptolemy's Geography of Eastern Asia*, 1909.

GEZELIUS (B.). *Japan i vesterlandska framstallning till om kring år 1700* (Linkjoping, 1910).

GILES (H. A.). *History of Chinese Literature*, 1927.

—— "Translations from the Chinese World Map of Father Ricci" (*R.G.S.J.*, Vol. LIII, 1919).

HAIFA MUNICIPALITY. *Old Maps of the Land of Israel Exhibition*, 1963.

HEAWOOD (E.). "The Relationships of the Ricci Maps" (*R.G.S.J.*, Vol. L, 1917).

IKEDA (Teturo). "World Maps in Japan before 1853" (*Congr. Int. Geogr.*, Amsterdam, 1938).

INDIA. *Catalogue of Maps, Plans, &c. of India and Burma and other parts of Asia*. Pub. by order H.M. Sect. State for India in Council, 1891.

KAMMERER (A.). *La découverte de la Chine par les Portugais au XVI siècle et la cartographie des Portulans* (Leiden, 1944).

KISH (G.). "Some Aspects of the Missionary Cartography of Japan during 16th Cent." (*Imago Mundi VI*, 1949).

KITAGAWA (Kay). "The Map of Hokkaido of G. de Angelis, c. 1621" (*Imago Mundi VII*, 1950).

LANZAS (P. T.). *Relación descriptiva de los Mapas, Planos, etc. de Filipinas* (Madrid, 1897).

MEYER (H. M. Z.). *Jerusalem: the Saga of the Holy City* (Jerusalem, Universitas-Pub., 1954).

NAKAMURA (Hiroshi). *Les Cartes du Japon qui servaient de modèle aux cartographes européens au début des relations de l'Occident avec le Japon* (Mon. Nipponica, 1939).

—— "Old Chinese World Maps, Preserved by Koreans" (*Imago Mundi IV*, 1947).

—— *History of Mapping of Eastern Asia* (Yokohama, 1958).

RAEMDONCK (J. van). "La Géographie Ancienne de la Palestine" (*Bull. de l'Acad. d'Archeol. de Belgique*, Ser. III).

RAMMING (M.). "The Evolution of Cartography in Japan" (*Imago Mundi II*, 1937).

ROHRICHT (R.). *Bibliotheca Geographica Palaestinae* (Berlin, 1890).

RUBENS (R.). *South East Asia, Maps of Borneo, Burma, Malaya, Thailand and Vietnam*. Map Collectors' Circle, 1975.

SIEBOLD (P. F. von). *Atlas von Land- und Seekarten von Japan* (Reich, 1851).

SOOTHILL (W. E.). "Two Oldest Maps of China Extant" (*R.G.S.J.*, Vol. LXIX, 1927).

TELEKI (P.). *Atlas zur geschichte der Kartographie der Japanischen inseln* (Budapest, Leipzig, 1909).

UHDEN (R.). "The Oldest Portuguese Chart of the Indian Ocean, 1509" (*Imago Mundi III*, 1939).

VACCA (G.). "Note sulla storia cartografia cinese" (*Riv. Geogr. italiana*, 1911).

VIVIEN DE ST. MARTIN (P.). *Etude sur la Géographie Grecque et Latine de l'Inde* (Paris, 1858).

CHAPTER XII

America

THE labour and munificence expended in the formation of collections of Americana, including cartography, and the care with which such collections have been examined have resulted in America being better documented than nearly any other land. Coupled with the fact that America has excited in one way or another the interest of the rest of the Occidental world during practically the whole of her history, with a consequently high output of printed works on the subject, the modern commentators' problem is one of selection rather than collection of data. So in a work of this nature it is only possible to mention a few of the more important or interesting items. More detailed information can be gathered by studying the works of the authorities given at the end of the chapter.

The oldest-recorded manuscript to date showing the discovery of the New World is Juan de la Cosa's map of 1500. This, the oldest source map known to us, is likewise the most correct in one important particular, showing a solid land barrier to expansion in the West, though it must be remembered that Columbus thought he had made his landfall on the eastern shores of Asia, and La Cosa was pilot to Columbus. Other important manuscript maps are the world portolans of Cantino and Caneiro, both compiled about 1502, and both Portuguese in origin, though the latter was a native of Genoa. Another map of great interest was the sketch-map of Bartholomew Columbus.

The first known printed map is the world map by Matteo Giovanni Contarini, engraved by Francesco Roselli in 1506 (15). The unique example is in the British Museum. In this map the discoveries in North America are shown as the eastern coast of a greatly extended Asian mainland, and a similar connection is indicated in the south. The West Indies are represented, but otherwise a wide-open ocean gap is shown between North and South America.

The much larger world map by Martin Waldseemüller, printed from twelve wood-blocks in 1507, is the second printed map to show America, and the first to so name it. This map shows a radical alteration from the Contarini conception. Its distinguishing feature is the long attenuated form given to America, the west coast of which is as it were rolled back to indicate its lack of knowledge of this area. The whole continent is separated from Asia by a wide gap, while the space between North and South America is contracted to a narrow strait (and in fact eliminated entirely in the smaller hemisphere above the main map). The northern coast terminates abruptly with open sea beyond, and the possibility of a passage round the southern end is evaded by continuing South America to the edge of the map. Waldseemüller's map is largely based on the Caneiro map of 1502 as regards outline, though being a land map the interior is filled in. Curiously enough Waldseemüller, while accepting the Portuguese delineation for the New World, reverts to the Ptolemaic conception for eastern Asia, rejecting the

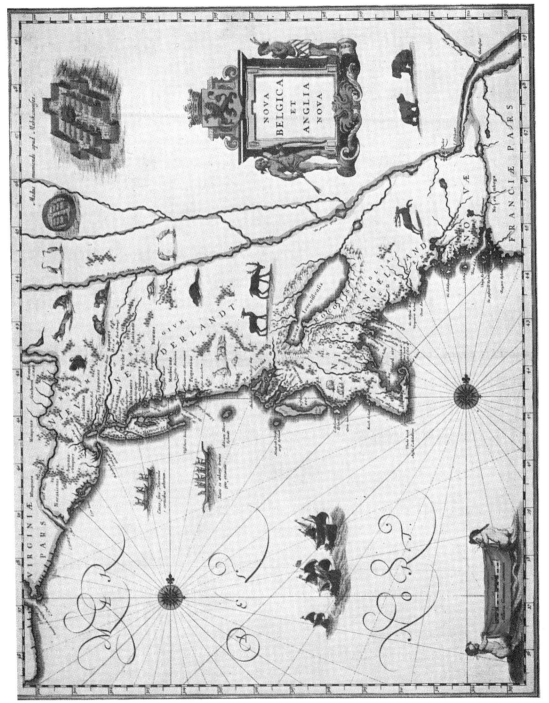

79 BLAEU'S NEW BELGIUM AND NEW ENGLAND (20" × 15")

from the *Atlas Major*, 1662

more correct rendering of Caneiro. Waldseemüller produced a second large-scale world map in 1516, the "Carta Marina." This was likewise printed on twelve sheets, and is an even closer copy of the Caneiro chart even down to the wind roses. India in this map is given a correct southern projection. The inland features are of course taken from other sources. Of a thousand copies known to have been printed, only one of each is known to survive. They were not located till 1901 by Prof. J. Fischer, in the collection of maps in Wolfegg Castle. A full-size reproduction of both maps was issued with descriptive text in 1903 by Fischer and Wieser.

The map of 1507 was copied in manuscript by Glareanus in 1510 and the small inset maps by Stobnicza in 1512. The large map was reproduced by Apian on a smaller scale in 1520, and as late as 1546 by Honter in his *Rudimenta Cosmographica*. It likewise influenced the world maps of Johann Ruysch, Schöner, Grynæus, Vadianus and Vopel.

The Ruysch world map, found in the 1508 edition of Ptolemy, is based on the Contarini map of 1506, even to its unusual fan-shaped projection, but with important changes, particularly in eastern Asia, where it more closely follows the Caneiro model. It was once thought to be the first printed map to show America; it is in fact, the first in an edition of Ptolemy.

The next important printed map to show America is the fine world map on a cordiform projection in the 1511 edition of Ptolemy printed in Venice. This heart-shaped map, with its surrounding wind cherubs, is a fine example of Italian workmanship of the period. In the same year Peter Martyr's *Decades* appeared, containing a map of the West Indies and the adjoining coasts. In 1513 the most valuable and important, after the first, of all the editions of Ptolemy appeared in Strassburg. Produced under the direction of Waldseemüller, it contains a world map known as the "Admiral's map" and a separate map of the known part of America—"Tabula Terre Nove"—one of the earliest printed maps devoted exclusively to the New World. Next came the Ruysch map in the 1515 edition of the Margarita *Philosophica Nova*, copied from the Strassburg Ptolemy for the American portion. In 1520 Peter Apian copied Waldseemüller's world map of 1507 for an edition of Solinus published in Vienna. Apian, however, leaves a passage round South America, though still showing an ocean strait between the two continents. This map reappeared in the 1522 edition of Pomponius Mela *De Orbis Situ*, published by Vadianus.

Various globes were likewise constructed before 1520; for example, those known as the Lennox (1510) and Boulenger (1514) and that of Schöner of 1515. Another much-publicised map is the world map of Laurent Frisius in the 1522 edition of Ptolemy.

In 1528 a world map on an oval projection was included in Bordone's *Isolario*. This work also contained some small maps of the West Indian islands and a plan of Mexico City. A similar but more detailed oval world map was engraved by F. Roselli for the *Isolario* of Bartolomeo dalli Sonetti.

In 1519, Oronce Finé, a French cartographer, had compiled a heart-shaped map of the world, and in 1531 completed a double heart-shaped projection with the title, "Nova et integra Universi Orbis Descriptio." This appeared in the work of Grynæus' *Novus Orbis Regionum* printed in Paris in 1532. It was re-issued in the 1540 edition of Pomponius Mela. The work of Grynæus was also issued in the same year in Basel with a fine map on an oval projection, the so-called Holbein World Map, one of the most charming pictorial maps of the period. The map is on the Waldseemüller model, with a narrow strait shown between North and South America, North America being marked "Terra de Cuba," and Cuba lettered "Isabella." South America, however, is considerably altered, being given a much greater width from east to west, tapering off to a long narrow peninsula in the south. The oval map is placed within a rectangular framework, the intervening space filled in with scenes of native life and customs in Asia and America, the map itself bearing ornamental plaques, magnificent dolphins, ship and mermaid. The numerous editions of Apian's *Cosmography* are of interest as including a representation of the world, including America, for example the editions of 1534, 1548, 1551, 1553, 1574, 1575 and 1584.

Two other maps to have long lives were the map of the world and the map of America in Münster's *Cosmography* (see Chapter IV). Münster's map of America is one of the quaintest of the 16th century, with the Portuguese standard off the coast of Africa, the Spanish standard in the West Indies, and a representation of Magellan's ship in the Pacific. South America is well defined with the estuary of the Plate, but North America has a west coast running due north and south, and Canada is roughly joined to New England by a narrow peninsula (80).

Finé's world map was utilised by Mercator in 1538, and Mercator's in its turn was copied by Antonio Salamanca in Rome. Another world map of Finé's was reproduced by Cimerlinus in 1566. Map production in Italy had by this time greatly increased. Gastaldi, the greatest of the Italian map-makers, produced a fine oval world map in 1546. This was exceedingly popular, and was copied by many of his contemporaries, for example by de Jode (1555), Forlani (1560 and 1562), F. Bertelli (1562 and 1565), L. Bertelli and D. Bertelli (1568), Camocio (1569), Duchetti (1570), Ortelius and Valgrisi.

In 1545 Medina's *Arte de nauegar* was printed in Valladolid. This work contained a map of the *Nuevo Mondo*, one of the very few printed maps from Spanish presses. It was re-issued with a slight addition in Seville in 1549, and in later editions of the work Nicolas de Nicolay's map was copied by Camocio, Forlani and Bertelli in Italy about 1560.

Other works to contain important maps were Ramusio's *Voyages*, which not only had a world map, but maps and plans of Mexico, Cusco, New France, Brazil, etc., and Girava's *Cosmography* of 1556, with a world map by Vopel: re-issued 1570.

A separate map of North America was issued by Zaltieri in 1566, though it was in fact derived from the western half of Gastaldi's world map, and a fine map of South America, "La Descrittione di tutto il Peru," of about the same date. Forlani also compiled a map of North and South America which was printed in 1574, and a large undated map of America on nine sheets was published by Bertelli and Camocio.

All the preceding maps are rare, though the great majority have been reproduced in one form or another. We now come to the standardised atlases of Ortelius (76) and Mercator, the first in 1570, the other in 1595. There are innumerable editions of these two atlases (see Chapter V). Both contain maps of America, that of Ortelius being derived from Gastaldi, and Mercator's from his large world map of 1564.

Both are extremely decorative, with large cartouches enclosing the title, with ships, sea monsters and other decorations. Both show California correctly as a peninsula, with the west coast stretching away north-westward, and with a great southern continent across the whole base of the map. The St. Lawrence is well delineated, but none of the Great Lakes is shown.

Towards the end of the century English cartographers made important contributions to the mapping of the New World, Humphrey Gilbert, Frobisher, Michael Lok, Robert Thorne, Drake, Ralegh, Molineux, William White: in this period also come the great collections of voyages by Hakluyt, De Bry and Linschoten (83), all of which have important maps.

In 1597 Wytfleet's *Supplement to Ptolemy's Geography* was printed. This was the first atlas to deal exclusively with America. The editions of this work are as follows:

1597 folio, Louvain, 19 maps.	1605 folio, Douay, 23 maps (4	1611 folio, Douay, 23 maps.
1598 folio, Louvain, 19 maps.	new maps on East Indies).	1615 folio, Arnhem, 19 maps.
1603 folio, Douay, 19 maps.	1607 folio, Douay, 23 maps.	

The 17th century saw a great increase in the number of maps devoted to America. In the early part of the century many of the most important maps first appeared in topographical works, either general or on particular areas, such as Wright's map in *Hakluyt* (1600), Smith's Map of Virginia (1612), Wood's New England (1634), the maps of Champlain (1606), and Lescarbot (1618), Tatton's California and Mexico (1616) (81), and general works, such as De Laet's *Nieuwe Wereldt* (1630) and Ogilby's *America* (1671) (87).

80 SEBASTIAN MÜNSTER'S "NEW WORLD" (13½" × 10")

from his *Cosmographiae Universalis* (Basle; 1540)

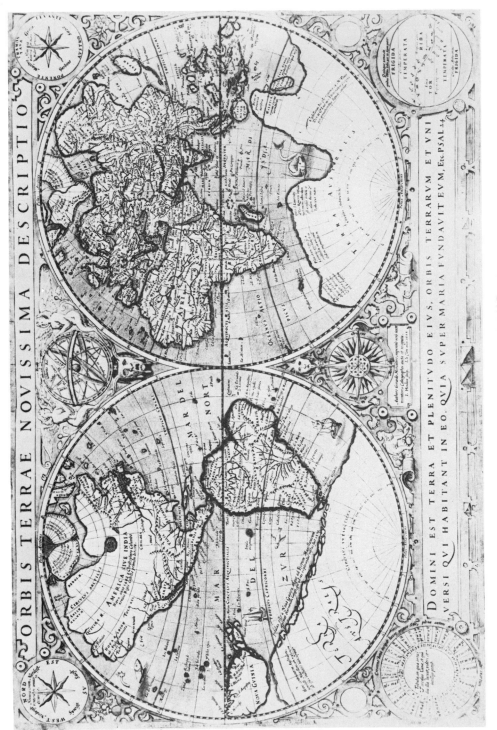

81 LE CLERC'S WORLD MAP, 1602

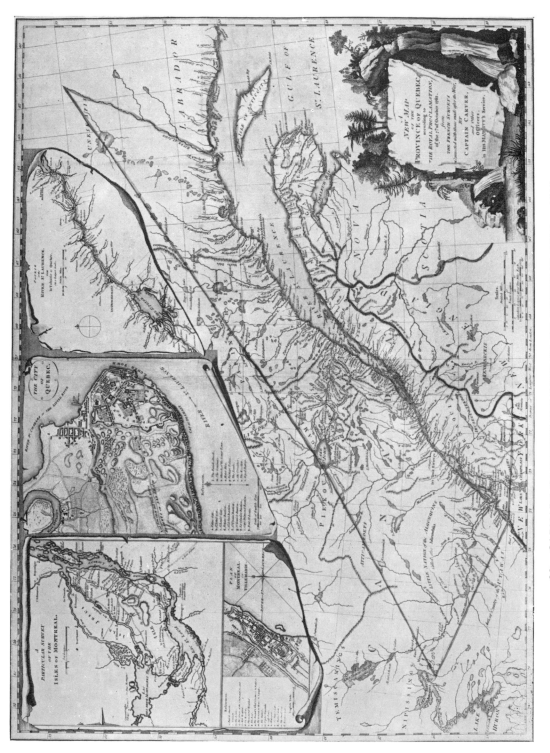

82   CAPTAIN CARVER'S "PROVINCE OF QUEBEC" ($26\frac{1}{2}" \times 19\frac{1}{2}"$)

from Jeffery's *American Atlas*, 1776

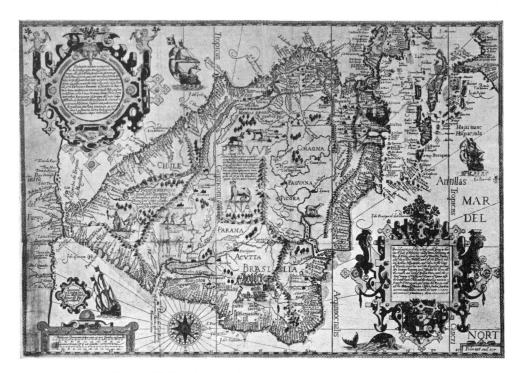

83   LINSCHOTEN'S SOUTH AMERICA (21″ × 15″)

from Linschoten's *Voyages*, 1599

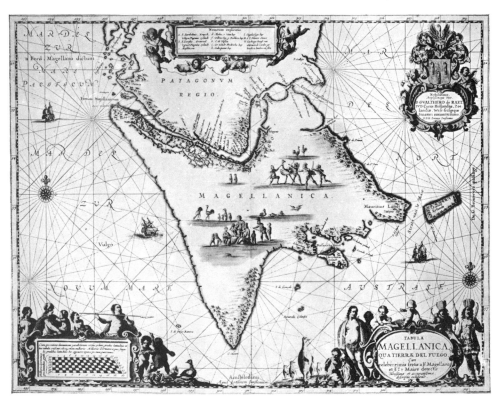

84   JANSSON'S *MAGELLANICA* (20½″ × 16″)

from *Novus Atlas*, 1658

The 17th century is the century *par excellence* of the decorative coloured map. Editions of Ortelius and Mercator's Atlases (the latter taken over by Hondius) continued to be printed, and they were followed by the atlases of Blaeu, Jansson (84), Visscher and De Wit, and Sanson (77), and the whole of the Dutch and French schools of this period (see Chapters V and VI for details of editions).

These atlases contained, not only general maps of North and South America, but separate maps of the various states. The first English general atlas, that by John Speed (1627), contained a general map of North and South America, with broad side borders containing large figures of the native inhabitants and a top border of small vignette bird's-eye views of the principal towns, and the same pattern was followed with greater elegance by Blaeu and Visscher. Speed's atlas also contained a separate map of the Bermudas. In the 1676 editions of his work, additional maps of New England and New York, Virginia and Carolina, Jamaica and Barbados were added. While not always the most advanced geographically, the most beautifully printed maps of the period are those by Blaeu. In them everything is on a grand scale, the paper superb, the calligraphy magnificent, the engraving clear, the decorations carefully executed, and in choice examples the colouring brilliant. Whether it is a map of New York (79), or the West Indies, of Argentina or Magellanica, the same high standard is maintained. No shoddy workmanship was allowed to come out of the house of the Blaeus. The maps of Jansson his rival come very close to, and in some cases are identical with, those of Blaeu. The early maps of Visscher are likewise very fine, but the later maps have coarsened somewhat in treatment, though of interest as showing advancement in geographical knowledge. The maps of the elder Sanson, the founder of the French school, are in quite a different style, with far less ornamentation than the Dutch school; they are nevertheless very attractive, their beauty lying largely in their fine lettering and careful spacing, their general effect being one of remarkable clarity. The engraving is delicate, great care being taken with the geographical symbols employed. Towards the end of the century the large maps of Jaillot, successor to Sanson, are remarkable examples of fine engraving with ornate title-pieces and magnificent when in full colour. Apart from the land maps, the sea atlases included many excellent charts of the American coasts, and it is always desirable to secure fine examples of odd plates from the marine atlases of Dudley, Seller (86), Blaeu, Colom, Doncker, Goos and Roggeveen when they occur for sale.

In the 18th century output of maps of America was extremely prolific, general and regional maps coming in great numbers as well as complete atlases mainly from the presses of England and France. The wars between England and France and later the War of Independence greatly stimulated the demand for new and correct surveys of large areas, and much fine work was done by military and naval engineers, the knowledge of these specialised maps being incorporated in the general maps issued to the public.

About 1710–15 Herman Moll issued a general atlas without a title containing 25 maps of large folio size, usually found folded in four, making a tall narrow volume. It was re-issued about 1730 with 30 maps. Moll's Atlas contains some popular maps, notable for the large vignettes they contain; for example, his map of South America has a large inset engraving of the Silver Mines of Potosi, and his North America an engraved representation of the various processes of the cod fishery. His map of "the Dominions of Great Britain in North America" is illustrated with notes on the natives, the fishery and the Post Roads, and includes a large vignette of Niagara, with beavers building a dam. His map of the "North Parts of America claimed by France" has a charming vignette of an Indian village. Similar maps, but without the vignettes, were issued by John Senex in the same period. Senex also issued some maps on a smaller scale in his *New General Atlas*, 1721, including maps of North and South America, English Empire in America, Virginia and Maryland, West Indies, Mississippi and Panama. All of these have charmingly designed title cartouches.

From France came maps by De l'Isle, Du Val, De Fer, Le Rouge, Janvier and D'Anville of the various parts of North and South America, and from Germany decorative maps by Bodenehr, Kohler, Homann and Seutter.

In 1733 an important atlas was issued—Henry Popple's *Map of the British Empire in America*—consisting of a key map and 20 sectional sheets. This was the best and most detailed map issued up to this date. Two important maps in the middle of the century were Lewis Evans' map of the Middle Colonies (1749) and John Mitchell's Map of the British and French Dominions in North America (1755). Both had a considerable influence on later maps, there being numerous re-issues and piracies. These have been exhaustively discussed by H. N. Stevens. In the latter part of the century other complete atlases devoted exclusively to America appeared, such as Thomas Jefferys' *American Atlas*. This was printed in 1776, and had 30 maps, including surveys by William Scull, Henry Mouzon, Captain Cook, Major Holland, Lewis Evans, W. Carver (82) and others. Jefferys also issued a *West India Atlas* in 1775, with 40 maps, a second edition appearing in 1780, also with 40 maps. Captain Joseph Speer produced his *West India Pilot* in 1766 with 13 maps, a second edition appearing in 1771 with the number of maps increased to 26. From France came the *Pilote Amériquain Septentrional* in 3 parts, 1778–9, with 54 maps (Part 3 with 13 maps being very rare).

William Faden, successor to Thomas Jefferys, was the greatest and most prolific map-maker at the end of the century. He published his *North American Atlas* in 1777. This had 34 maps. He likewise issued American atlases without printed titles, and many of these contain fine large-scale regional maps and magnificent plans, such as Blaskowitz's Rhode Island, Brahm's Carolina, Montresor's New York, plans of Boston by Lieut. Page, New York by Ratzer, Philadelphia by Easburn and by Scull and Heap. Many of the military surveys and plans of military operations in the War of Independence were magnificently engraved, and form some of the finest examples of map-making in the 18th century.

The other great work of this period was the *Atlantic Neptune*, compiled mainly by J. F. W. Des Barres for the Admiralty. Des Barres, originally Swiss, became a British subject and entered the Royal Military College at Woolwich. He accompanied Wolfe in the Quebec Campaign as Engineer. Des Barres gradually compiled his charts during his service under the Crown, and they were issued in atlas form to various commanders as required. No two copies are alike. Engraving started about 1774, but it was not till 1777 that the charts began to appear under the title *Atlantic Neptune*. Des Barres' charts are remarkable for the wealth of topographical detail, a rare feature in a marine map. Many place-names are shown that appear on no other maps of the period, and owing to changes in importance and nomenclature, are in some cases otherwise unknown: thus they are extremely valuable historically. Des Barres supplemented his charts with a series of views, many of these beautifully engraved and coloured like water-colours. Some were engraved as insets on the charts, but the more important were issued as separate plates. The value of such collections depends largely on the number of the views included, some of which are exceedingly rare.

At the turn of the century the large-scale maps of Aaron Arrowsmith must be mentioned. He issued the following:

North America: 1795, 1811 and 1815. United States: 1796, 1802, 1808, 1815 and 1819; French edition 1802. West Indies: 1803. North and South America: 1804 and 1808. Panama: 1800. Buenos Aires: 1806. South America: 1811, 1814 and 1817.

In the 19th century, native cartographers took the place of European map-makers. Carey and Lea produced an American atlas in 1832, and Tanner a general atlas in 1840. Henceforward American map-makers—a list of whose works falls beyond the scope of this work—supplied the main contribution to the mapping of the country.

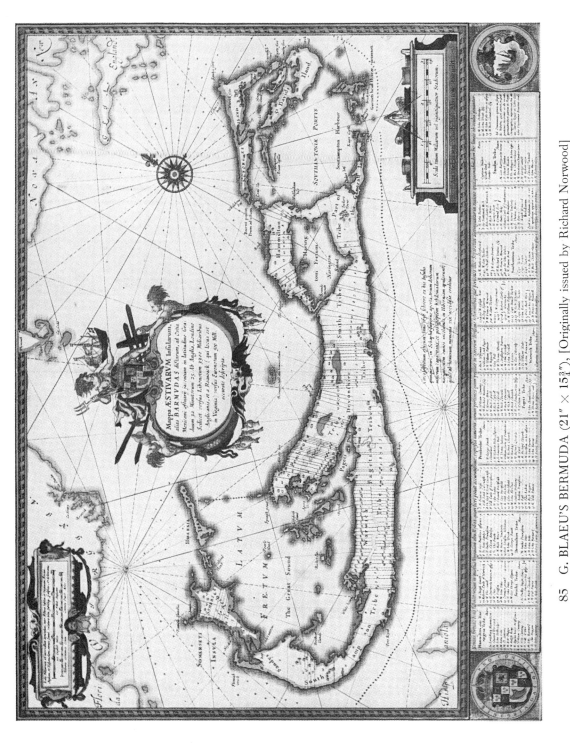

85   G. BLAEU'S BERMUDA (21″ × 15¾″), [Originally issued by Richard Norwood]
from the *Atlas Major*, 1662

AUTHORITIES

ADAMS (R. G.). *British Headqutr. Maps and Sketches used by Sir Henry Clinton . . descriptive List* (William L. Clements, Lib., Ann Arbor, 1928).

ALBA (Duque de). *Mapas españoles de América, sig.XV-XVII* (Madrid, 1951).

ANDREWS (W. L.). *New Amsterdam, New Orange, New York* (N.Y., 1897).

ASHER (G. M.). *Bib. and Hist. Essay on Dutch Books and Pamphlets relating to New Netherland . . . as also on the Maps, Charts, &c.* (Amsterdam, 1854–67).

BANCROFT LIBRARY, SAN FRANCISCO. *Index to Printed Maps* (Boston, Mass., 1964).

BEAGLEHOLE (J. C.). *Exploration of the Pacific* (London, 1966).

BEANS (G. H.). *Some 16th-Century Watermarks found in Maps prevalent in the Iato Atlases*, 4to, 1938.

BEAZLEY (C. R.). *John and Sebastian Cabot* (New York, 1898).

BOSTON. "List of Maps of Boston pub. between 1600–1903" (Boston Municipal Printing Office, 1903).

BRITISH MUSEUM. Contarini's Map of the World, 1924.

—— Sir Francis Drake's Voyage round the World, 2 contemporary maps, folio, 1931.

BROWN (Lloyd A.). *The Story of Maps* (Boston, 1949).

CANADA. Cat. of Maps and Charts in the Map Room of the Dominion Archives, Ottawa, 1912.

—— Catalogue of maps of the Geographic Board (Ottawa, 1922).

—— Map Collection of Public Reference Library, City of Toronto, 1923.

CHRISTY (Miller). *The Silver Map of the World*, 8vo, 1900.

—— *On an Early Chart of the N. Atlantic*, in Royal Library, Stockholm, 8vo, 1897.

COSA (Juan de la). *Mapa Mundi. Ensayo Biográfico del célebre Navegante . . . e historia de su famosa Carta Geográfica por Ant. Vascano*, 4to, Madrid, 1892.

—— *The Mappemonde of Juan de la Cosa*, by G. E. Nunn (1935).

CUBA. *List of Books relating to Cuba, by A. P. C. Griffin, with Bibliography of Maps by P. Lee Phillips* (Library of Congress, Washington, 1898).

CUMMING (W. P.). *The South East in Early Maps* (Princeton, 1958).

—— *North Carolina in Maps* (Raleigh State Dept. of Archives & History).

D'AVEZAC. *Les Voyages d'Améric Vespuce au compte de L'Espagne* (Paris, 1858).

—— *Martin Hylacomylus Waldseemüller: ses ouvrages et ses collaborateurs* (Paris, 1867).

DAWSON (S. E.). *Memorandum upon the Cabot Map*, 8vo, Ottawa, 1898.

—— *The Discovery of America by John Cabot in 1497 and the Voyages of the Cabots*, 3 parts, 8vo (Trans. of Roy. Soc. of Canada, Ottawa, 1896–7).

DAY (James). *Maps of Texas 1827–1900* (Austin Pemberton Press, 1964).

ELTER (A.). "De Henrico Glareano" (*Festschrift der Univ. Bonn*, 1896).

FISCHER (J.). *Discoveries of the Norsemen in America*, roy. 8vo, 1902.

—— and Prof. F. R. v. WIESER. *Facsimiles of the Waldseemüller World Maps of 1507 and 1516*, 1903.

FITE (E. D.) and A. FREEMAN. *A Book of Old Maps Delineating American History*, folio, 1926.

GALLOIS (L.). "Améric Vespuce et les géographes de S. Dié" (*Bull. Soc. Géogr. de l'Est*, 1900).

GANONG (W. F.). *A monography of the cartography of Province of New Brunswick* (Trans. R. Soc. Canada, 1897).

GRAVIER (G.). *Carte des Grands Lacs de l'Amérique du Nord*, 4to, Rouen, 1895.

—— *Etude sur une Carte inconnue* (Joliet's Mississippi), 4to, Paris, 1880.

—— *Etude sur le Sauvage du Brésil* (Paris, 1881).

GRINO (Prof. S.). *Schizzi cartografici inediti ui primi anni della scoperta dell' America*, 8vo, 1930.

GUILLEN Y TATO (J. F.). *Monumenta Chartographica Indiana*, 1942.

GUTHORN (P. J.). *American Maps and Mapmakers of the Revolution* (Freneau Press, 1966).

GUZMAN (Prof. E.). "The Art of Map-making among the Ancient Mexicans" (*Imago Mundi III*, 1939).

HARRISSE (H.). *Notes pour servir à l'histoire, à la Bibliographie et à la Cartographie de la Nouvelle France, 1545–1700*, 8vo, Paris, 1872.

—— *Discovery of North America . . . with an Essay on the Early Cartography of the New World*, 3 vols. 4to, 1892.

—— *Découverte et évolution cartographique de Terre-Neuve, 1479–1769*, 4to (London, 1900).

HARRISSE (H.). *Bibliotheca Americana vetustissima* (New York, 1866. Additions 1872).

HEAWOOD (E.). "Jodocus Hondius Map of the World on Mercator's Projection, 1608" (*R.G.S.*).

HOBBS (W. H.). "Zeno and the Cartography of Greenland" (*Imago Mundi VI*, 1949).

HOLMDEN (H. R.). *Catalogue of Maps, Plans and Charts in the Map Room of the Dominion Archives*, 8vo, Ottawa, 1912.

HULBERT (A. B.). *Crown Collect. of Photographs of American Maps* (New York, 1912).

HUMPHREYS (A. L.). *Old Decorative Maps and Charts*, 4to, 1926.

JOMARD (E. F.). *Les Monuments de la Géographie*, folio atlas, with 8vo text (Paris, 1854–6).

KARPINSKI (L. C.). *Bibliography of the Printed Maps of Michegan, 1804–1880*, 8vo, 1931.

KIMBALE (L. E.). "James Wilson of Vermont, America's First Globe Maker" (*Proc. Am. Antiq. Soc.*, 1938).

KIMBLE (G. H.). *Catalan World Map of R. Biblioteca Estense at Modena* (1933).

KOHL (J. G.). *Descriptive Catalogue of those Maps, Charts, and Surveys relating to America which are mentioned in Vol. III of Hakluyt* (Washington, 1857).

—— *Die Beiden Ältesten General-Karten von Amerika, 1527 und 1529*, folio (Weimar, 1860).

KOHLIN (H.). "First Maps of Delaware, a Swedish Colony in N. America" (*Imago Mundi V*, 1948).

KRETSCHMER (K.). *Die Entdeckung Amerikas in ihrer Bedeutung für die Geschichte des Weltbildes*, folio text and atlas (Berlin, 1892).

KUNSTMANN (F.). *Die Entdeckung Amerikas nach den ältesten Quellen geschichtlich dargestellt*, 4to text and folio atlas (München, 1859).

LA COSTA (Juan de). *Portolan World Chart*, A.D. *1500* (Madrid, 1892).

LE GEAR (Clara Egli). *United States Atlases. A List of National, State, County, City and Regional Atlases in the Library of Congress*, Washington, 1950.

LOWERY (Woodbury). *The Lowery Collection, Descriptive List of Maps of the Spanish Possessions within the present limits of U.S.* (1912).

LUCAS (F. W.). *Appendiculae Historicae, or Shreds of History hung on a Horn*, 4to, 1891.

—— *The Zeno Question*, 4to, 1898.

LUNNY (R. M.). *Early Maps of North America* (Newark, N.J., 1961).

MACDONALD (Dr. A. E.). *Early Maps of Canada 1511–1886* (Ontario, 1958).

MACFADDEN (C. H.). *A Bibliography of Pacific Area Maps* (Modern American Council. Inst. of Pacific Relations, 1941).

MAJOR (R. H.). *Memoir of a Mappemonde by Leonardo da Vinci*, 4to, 1865.

—— *The True Date of the English Discovery of America under John and Sebastian Cabot*, 4to, 1870.

MARCEL (G.). *Cartographie de la Nouvelle France, supplément à l'ouvrage de M. Harrisse*, 8vo, Paris, 1885.

—— *Reproduction de Cartes et Globes relatives à la Découverte de l'Amérique*, 4to text, folio atlas (Paris, 1893–4).

MERCATOR. *Drei Karten von Gerhard Mercator Europa, Britische Inseln, Weltkarte*, folio, Berlin, 1891.

NORDENSKIÖLD (A. E.). *Facsimile Atlas*, folio, 1889.

—— *Periplus*, folio, 1897.

NUNN (G. E.). *The Geographical Conception of Columbus*, 8vo, New York, 1924.

—— *World Map of Francesco Roselli*, 4to, 1928.

—— *Origin of the Strait of Anian Concept*, 8vo, 1929.

—— *The Columbus and Magellan Concepts of South American Geography* (1932).

—— *The Mappemonde of Juan de la Cosa, a critical investigation of its date* (1934).

—— *Antonio Salamanca's Version of Mercator's World Map of 1538* (1934).

OBER (F. A.). *John and Sebastian Cabot* (New York, 1908).

OBERHUMMER (E.). *Zwei handschriftliche Karten des Glareanus* (Munchen, 1892).

PALMER (M.). *Printed Maps of Bermuda* (London, Map Collectors' Circle, 1965).

PHILLIPS (P. Lee). *List of Geographical Atlases in the Library of Congress, with Bibliographical Notes*, 4 vols., 4to, 1909–20.

—— *Alaska and the N.W. Part of N. America, 1588–1898. Maps in the Library of Congress*, Washington, 1898.

—— *List of Maps and Views of Washington and District of Columbia in Library of Congress*, Washington, 1900.

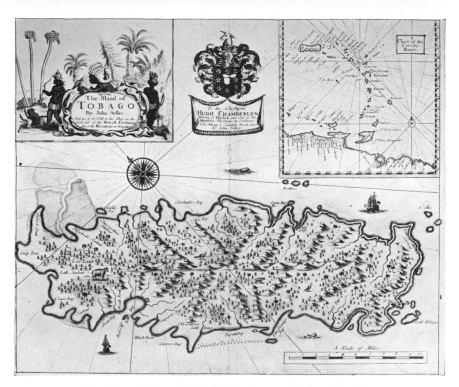

86   JOHN SELLER'S TOBAGO, *c.* 1660 (20″ × 17″)
(*B.M. Maps* 82510(4))

87   JOHN OGILBY'S VIRGINIA (15″ × 11½″)
from Ogilby's *America* 1671

88   TALLIS'S MEXICO, CALIFORNIA AND TEXAS ($11\frac{3}{4}'' \times 8\frac{3}{4}''$)

from *The Illustrated Atlas, c.* 1851

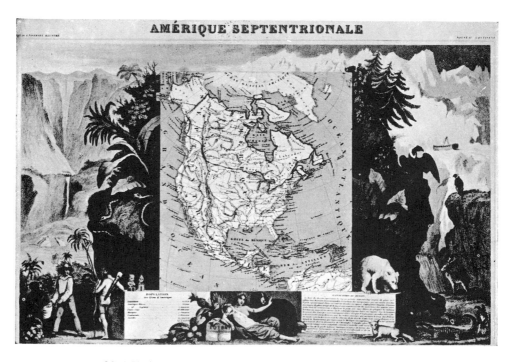

89 LEVASSEUR (V.) NORTH AMERICA ($17'' \times 11''$)

from *Atlas National* (Paris, Combette, 1847)

RAVENSTEIN (E. J.). *Martin Behaim, his Life and his Globe*, 4to, 1908.

Remarkable Maps of the 15th, 16th and 17th Centuries, 6 parts, folio, 1894–7.

REPS (J. W.). *The Making of Urban America* (Princeton, 1965).

RHODE Is. Chapin (H. M.). *Check List of the Maps of Rhode Island, Providence*, 1918.

RISTOW (W. W.). "Seventeenth Century Wall Maps of America and Africa" (*Quarterly Journal of Library of Congress*, 1967).

RONCIERE (M. Ch. de la). *Un Atlas inconnu de la dernière expédition de Drake*, 8vo, 1909.

—— "Manuscript Charts of John Thornton" (*Imago Mundi XIX*, 1965).

SANTAREM (Vicomte de). *Atlas composé de Mappemonde et Portulans … VI–XVII Siècle*, 78 fol., 1842–53.

—— *Recherches sur Améric Vespuce* (1836).

—— *Memoirs sur les Colonies Angloises* (1840).

SHILSTONE (E. M.). "List of Maps of Barbados" (*Jnl. Barbados Museum and Hist. Soc.*, 1938).

SKELTON (R. A.). MARSTON (T. E.) and PAINTER (G. D.). *The Vinland Map and the Tartar Relation* (Yale U.P., 1965).

SMITH (Clara). *Newberry Library: List of MS. Maps in Edward E. Ayer Collection* (Chicago, 1927).

—— (Priscilla). *Early Maps of Carolina and adjoining Regions from collection of Henry P. Kendall*, ed. L. C. Karpinski, 8vo, 1930.

STEVENS (H.). *Hist. and Geogr. Notes on the Earliest Discoveries in America, 1453–1530*, 8vo, 1869.

—— *Johann Schöner of Nuremberg*, 8vo, 1888.

STEVENS (H. N.). *The First Delineation of the New World and the first use of the name America on a Printed Map*, 4to, 1928.

—— *Lewis Evans, his Map of the Middle British Colonies in America*, 4to, 1924.

STEVENS (Henry). *Notes Biographical and Bibliographical on the Atlantic Neptune*, 1937.

STEVENSON (E. L.). *Maps illustrating Early Discovery and Exploration in America* (1903–6).

—— "Typical Early Maps of the New World" (*Bull. Amer. Geogr. Soc.*, 1907).

—— *Marine World Chart, 1502, by Nicolo de Caneiro Januensis* (New York, 1907).

—— *Early Spanish Cartography in the New World* (Worcester, Mass., 1909).

—— *A Description of Early Maps, Originals and Facsimiles, 1452–1611*, being part of the permanent wall exhibition of Amer. Geogr. Soc. (N.Y., 1921).

STOKES (J. N. Phelps). *The Iconography of Manhattan Island*, 4to, 6 vols., New York, 1915–1928.

THACKER (J. B.). *Continent of America, its Discovery, its Baptism* (N.Y., 1896).

THOMASSY (B.). *Cartographie de la Louisiane*, 4to, Nouvelle Orleans, 1859.

THOMPSON (E.). *Maps of Connecticut before 1800* (Windham, 1940).

TIELE (P. A.). *Mémoire Bibliographique sur les Journaux des Navigateurs Néerlandais*, 8vo, 1867.

TOOLEY (R. V.). *California as an Island* (London, Map Collectors' Circle, 1964).

—— *French Mapping of the Americas* (De L'Isle 1966).

—— *Maps of West Indies, Trinidad and Tobago, 1964, Antigua 1969, Dominica and Grenada, 1970.*

—— *A Sequence of Maps of America, 1973.*

—— *Printed Maps of America*, Parts I–IV, A-Bent, 1970–73. All published.

VERNER (Coolie). *Carto-bibliographical Study of the English Pilot: the Fourth Book* (Charlottesville Virginia, 1960).

—— *Maps of the Yorktown Campaign 1780–81* (London, Map Collectors' Circle, 1965).

VINDEL (F.). *Mapas de América en los libros españoles* (Madrid, 1955).

WAGNER (H. R.). *Cart. of the N. W. Coast of America to year 1800*, 2 vols., Berkeley, California, 1937.

WHEAT (Carl I.). *Mapping the American West 1540 1857*, 5 vols., Worc., Mass., 1954.

WIESER (Dr. F.). *Magalhaes-Strasse und Austral-continent aus den Globen des Johannes Schöner* (1881).

—— *The Oldest Map with the name America* (Innsbruck, 1903).

WINSHIP (G. P.). *Cabot Bibliography*, 8vo, 1900.

WINSOR (Justin). *Narrative and Critical History of America*, 8 vols., 4to, Boston and New York, 1889.

—— "Baptista Agnese and American cartography in 16th Cent." (*Mass. Hist. Soc.*, 1897).

—— *The Kohl Collection of Maps relating to America* (Camb., Mass., 1886).

WROTH (C.). *Early Cartography of the Pacific* (New York, 1944).

CHAPTER XIII

Australia

AUSTRALIA provides one of the most fascinating of all studies in regional geography. Even as its fauna possesses many unique features, so a study of its historical cartography reveals peculiarities that can be found in no other series of maps.

Alone among the continents of the world, Australia was in a sense imagined centuries prior to its actual discovery. Pre-Christian cosmographers from a known land-mass in the north postulated a great southern continent to counterpoise the globe. In this global conception they were followed by the mediævals, and this geographical theory persisted down through the centuries, gradually shrinking in size as the voyages of successive navigators pushed its bounds ever farther to the south, but never totally destroyed till the days of Captain Cook. In fact, with the rounding of the Cape of Good Hope and the discovery of America, the belief in a great south land ceased to be a speculation merely for scholars, but was eagerly sought by adventurers, who hoped to find another El Dorado. It is true that certain conservative or cautious geographers in the 16th century and later (Edward Wright among others) did not insert this land on their maps, though that in itself did not imply a disbelief in its existence. But such maps form a small minority, and the most popular and famous world maps, composed by the leaders in cartographical thought and enterprise from the great Mercator himself, Gastaldi, Ortelius, their successors and imitators, show a continuous land-mass stretching across the base of their maps, and this cartographical figment exists in varying forms right down to the 18th century, and was finally dispelled by Captain Cook.

Marco Polo, in his travels in the 14th century, had spoken of three lands in the far south, Beach, Luchach and Maletur. Map-makers of the 16th century showed an upward projection of the southern continent curiously enough in the approximate position of Australia, and to this projection they gave the names cited by Marco Polo. Some writers have attempted to prove from this a European knowledge of Australia prior to the discovery of the Dutch and Spanish in 1605. This has no foundation in any fact so far revealed. History can be just as distorted as geography once fact is abandoned and theory advanced, and a treatise composed as fantastic in its way as the assumptions of certain early geographers of these regions. It should not be forgotten that the early cartographers put into their maps what they wished to believe. Thus a strait was shown at first between North and South America because it was desired, and a north-west and a south-west passage shown long before their actual discovery for the same reason. These limitations accepted, it is quite justifiable to include one or two early world maps in an Australian collection.

Another peculiarity of Australian maps is that, although Australia was discovered in 1605, there are no maps of Australia as such, generally speaking, until the start of the 19th century.

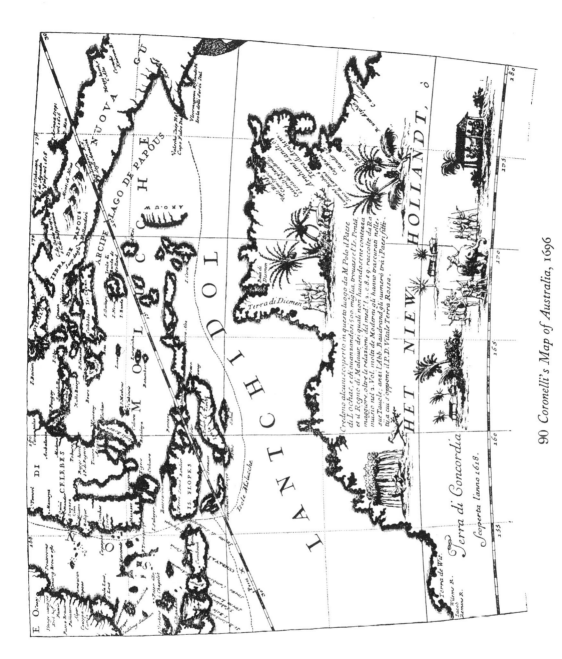

90 *Coronelli's Map of Australia, 1696*

Early representations of the Australian coast must be sought for amongst maps of the world, Asia, the East Indies, the Pacific, the South Pole, and even, in the case of New Zealand, in maps of America, or rather the Western hemisphere. For roughly 200 years after its discovery no inland features were shown in the maps of Australia, and no other land has been mapped so variously or erroneously. For a long time New Guinea was thought to be part of the Australian mainland. To this French geographers added a large tract in the north-east, of the discoveries of Quiros greatly enlarged. Tasmania was shown joined to Australia at times by an actual, at others by a hypothetical, line up to the time of Bass. New Zealand has been shown as part of Australia, and as joined to the Antarctic continent.

Even its nomenclature has not been static. The earliest names applied to this southern land-mass were the Land of Beach, Patalie Regio (Nether Regions), and Terra Australis Incognita. To the early Dutch it was the Land of Eendracht, and Land of the South. After the discovery of Tasman, it was named Company's New Netherlands. This was quickly changed to New Holland. In 1780 Daniel Djurberg named the continent Ulimaroa. William Faden in 1810 was one of the first to revert to the older form and name the continent Australia.

The Spanish expedition from the west, in which Torres found his strait, was not recorded except in manuscript form, and had no effect on the mapping of this part of the world. On November 28th, 1605, a Dutch ship, the *Duyfkin* (or Dove) entered the Gulf of Carpentaria. Other expeditions followed: in 1616 (the Eendracht), 1618 (the Mauritius), 1619 (Edel), 1622 (the Leuwin), 1628 (De Witt) and 1629 (Pelsart).

It was, however, some years before the first of these discoveries appeared on printed maps. Tasman was the first to alter radically the mapping of these regions. In his voyages he connected up the previous Dutch explorations into a connected whole, and added a knowledge of Tasmania and part of New Zealand, and his map became the standard for these regions up to the time of Captain Cook. Cook filled in the east coast and the whole of New Zealand in 1769–71, and Bass, King and Flinders filled in the details from 1797 to 1822.

Exploration of the interior was not seriously undertaken till the 19th century. Wentworth crossed the Blue Mountains in 1813. In 1814 Evans discovered the Lachlan and Macquarie Rivers. Sturt's expedition started in 1828 and added a knowledge of the Darling and Murray Rivers. Leichardt (1848), Burke and Wills (1860), M'Douall Stuart, (1858–62) and John Forrest (1868–74) in Western Australia, Ernest Giles (1872–6) in northern Australia, together with the spread of the inhabited parts, enabled the interior to be mapped gradually. Many of the maps of this period being undated, it is useful to remember the dates of the incorporation of the various states:

New South Wales.	1787 First settlement went out.
	1788 Sydney Cove.
Tasmania.	1642 Tasman.
	1798 Bass and Flinders.
	1803 British Possession.
	1825 Separated from Government of N.S.W.
South Australia.	1834 Created British Possession.
	1836 Adelaide founded.
	1842 Crown Colony.
Western Australia.	1697 Swan River—name given by Vlaming.
	1824 Brisbane founded.
	1829 British Possession.
Victoria	1803 Flinders surveyed coast.
	1836 Named Australia Felix by Sir T. Mitchell.
	Country round Port Philip occupied.
	1838 Melbourne.
Queensland.	1859 State formed.

91   ROBERT DUDLEY'S CHART OF NEW GUINEA AND PART OF AUSTRALIA
(15″ × 18½″)
from *Arcano del Mare*, 1646

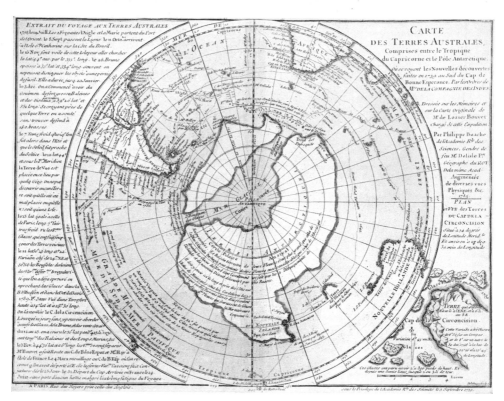

92   P. BAUCHE'S *TERRES AUSTRALES*, WITH NEW ZEALAND AS PART OF
THE ANTARCTIC CONTINENT (12¼″ × 9½″), 1754

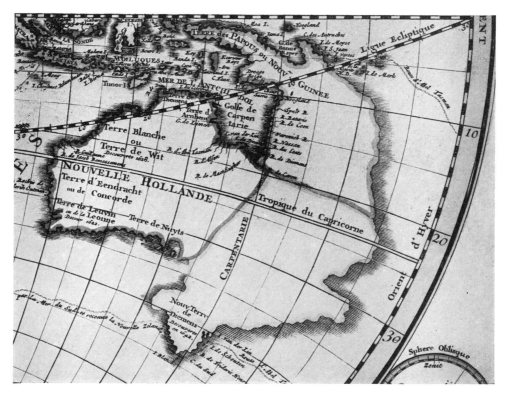

93   AUSTRALIA AS SHOWN ON SANSON'S WORLD MAP (COVENS AND
MORTIER *c.* 1720)

New Zealand.   1642 Discovered Tasman.
1769–70 Surveyed Captain Cook.
1814 Christian missionaries.
1839 N.Z. Company for colonisation.
1840 Sovereignty proclaimed over North Island by Captain Hobson.

Pre-discovery World Maps showing a Great Southern Continent

De Jode, 1593. Finé, 1531. Frobisher, 1578. Gastaldi, 1581?. Girava, 1556. Hondius, 1637. La Popelinière, 1582. Linschoten (1598). Mercator, 1538 and 1595. Münster, 1540. Myritius, 1590. Ortelius, 1570. Porcacchi, 1572. Purchas, 1625. Plancius, 1595. Quad, 1608. Schöner, 1520. Vopel, 1543. Wytfliet, 1597.

Asia showing Australia Pars in bottom right-hand corner separated from New Guinea by a narrow strait.

Ortelius, 1570. Mercator, 1595. Hondius, 1603.

East Indies, 16th Century

Ortelius. Doubt expressed as to the insularity of New Guinea. Beach pars continentis Australis shown below Java.
Visscher, 1617. Moluccas. Similar representation.
De Jode, 1593. New Guinea. Large exotic land below New Guinea.

Early Dutch—Pre-Tasman

Hondius. World, 1630–3. Shows west coast of Carpentaria with 8 names.
Ecckebrecht. World, 1630–3. In addition to above, shows part of the west coast of Australia with 6 names. The first printed map to show the west coast. Rare.
Jansson. East Indies (1633). Duyfkin's Eyland. Lat. 4° 13′ 3″.
Piscator (i.e. Visscher). World, 1639. Carpentaria with 8 names, farther to west Beach prov. aurifera below Java.
Hondius. E. Indies (1640). 't lant van de Eendracht and G. F. de Wits land.
Blaeu. E. Indies (1640). Carpentaria with 9 names and part of north-west coast of Australia with 4 names. An important map, the best to appear to date.
Dudley. "Carta particolare della costa Australe scoperta dall' Olandesi," Asia Carta XVIII. Engraved Lucini. This is the first separately printed map of Australia. It depicts Carpentaria and gives 24 names. Torres Strait lettered "Golfo Incognito." From the very rare sea atlas, "Arcano del Mare," of Robert Dudley (91).
Sanson-Mariette. World, 1651. Curious configuration to Australia showing south and west coast, but no north coast: "Beach" only.
Piscator (i.e. Visscher), 1652. Retrograde step, great southern continent shown with Beach below Java.
Jansson. East Indies, 1652. Australia called, "Terra del Zur," 9 names in Carpentaria, and coast from De Wit's Land to Nuyts Land. From Jansson's rare marine atlas.
Similar maps were issued by Colom, Goos and De Wit. Sanson's Moluccas, 1654, shows the discoveries in Carpentaria.

Early Dutch—Tasman's Period.

Alphen. East Indies (1660). One of the first charts to show Tasman's Discoveries, Tasmania, New Zealand, etc. From Alphen's Zee Atlas. Rare.
De Wit. World, 1660. Australia and New Zealand according to Tasman.
Van Loon. East Indies (1660). Australia according to Tasman with 34 names.
De Wit. East Indies, 1662. North coast of Australia after Tasman.
Thevenot. (Australia, 1663). The First French map of Australia from Thevenot's *Relation de divers voyages*. Tasman's representation copied from Blaeu's map of the East Indies of 1659. In the first issue of Thevenot's map the Tropic of Capricorn is not drawn on the plate. This map was copied by Bowen in 1747.

Other cartographers who gave Tasman's representation for this area were Roggeveen (1676), Allard (1676), Danckerts (1680), Van Keulen (1680) and Mortier (1700), also shown on charts of the East Indies by Hondius-Jansson, Doncker, Goos, Visscher and Robijn. All these maps based on Tasman were founded on actual observation, and where knowledge ceased the map was left blank. They were sound and reliable maps.

French School—Tasman plus theory

As the 17th century advanced, French geographers and map-makers came more and more to the front. They adopted Tasman's map as far as it went, but impatient for completion they attempted to fill in the gaps left in his map by theory. This retrograde step led to the most misleading and fantastic maps ever compiled for this region.

French School

From the very first, French representation of this area was peculiar. On Sanson the elder's world map of 1651 the west and south coasts of Australia are shown reasonably well, but the north coast is entirely imaginary, being based on the legendary configuration of the 16th century, with the "Beach" of Marco Polo. A few years later Nicolosius gave the same delineation, but with the addition of Van Diemen's Land and Arnhem's Landt. This last is correctly drawn, but removed from Australia, two northern coasts being shown on the same map.

A distinctive feature of the later French school was the insertion, greatly enlarged, of the discovery of Quiros. This was marked "Terra de Quiros" or "Terre Australe du Saint Esprit." By joining up Australia, New Guinea, Terra de Quiros and Tasmania, one vast continent was formed. This representation appears on the world maps of Du Val (1676), Sanson-Jaillot (1691), Nolin (1705), De Lisle (1714), Renard (1715), Buache (1741), Vaugondy (1752), Bellin (1753), Janvier (1760), Denis (1764), and Le Rouge (1778). A good example of this class is the large Sanson-Jaillot World Map (93). Australia, being normal as far as Carpentaria, is then extended as far again to the east; Tasmania, greatly enlarged and too far to the south, is joined in one continuous line to New Guinea in the north.

New Zealand was likewise given a theoretical form, a hypothetical east coast being added to the discovery of Tasman stretching away to the coast of South America. This is shown on the world maps of Sanson-Rossi (1674), Du Val (1676 and 1686), Coronelli (1695), Sanson-Jaillot (1700), and Nolin (1705).

New Zealand was also shown on various maps of the Southern Hemisphere removed from the Australian orbit and shown as a projection of a compact Antarctic continent, for example on the map of Buache (1754) (92), and Moithey (1760). It may be mentioned in passing that New Zealand is also shown on some maps of America, for example, Janvier's "Western Hemisphere" (1783).

In fairness to French publishers, it must be stated that these erroneous conceptions were corrected or discarded in later editions of their works. A commendable feature was the insertion of the tracks of celebrated navigators, Tasman, Dampier, Le Maire, etc.

Late Dutch School under French Influence

The decline of the Dutch map-makers after 1700 is strikingly emphasised in the cartography of Australia. Discarding the exact charting of their predecessors, they merely reproduced French theoretical geography, for example, Allard (1710), De Leth (1730), Ottens (1740), Ewyk (1750), Tirion (1753).

Australia Joined to New Zealand

A typical example is the strikingly decorative but fantastic map of the South Seas by Henry de Leth. Based on the map of Guedeville, it depicts the world from Australia to the

94 JOHN OVERTON'S, MAP OF AMERICA 1668

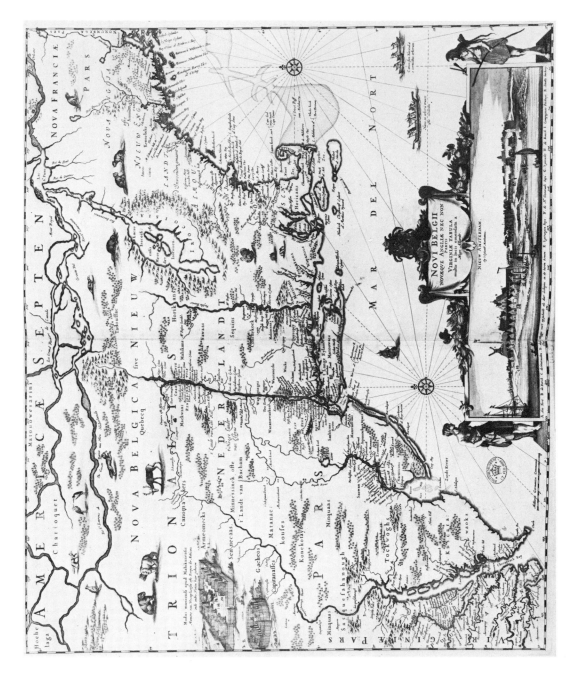

95 H. ALLARD'S NOVI BELGII (NEW YORK AND NEW ENGLAND)
1656. FIRST STATE

west coast of Africa, and has many inset plans and small vignette views and figures. The western part of Australia is shown approximately correct, but the south coast is continued eastward and joined on to a greatly enlarged New Zealand, with Terre de Quir forming an eastern coast. Tasmania is shown separated from the continent.

Another extraordinary map as coming from a Dutch Press is the World Map of Tirion (1753), from which Tasmania and New Zealand are omitted entirely.

A further curiosity is Covens and Mortier's map of the East Indies, on which New Guinea is shown twice, under its own name and farther to the west as Terra de Papous.

Early English Map-makers

The first English map to depict part of Australia is that in Dudley's rare work, the *Arcano del Mare* (1646): "Carta Particolare della costa Australe scoperta dall Olandesi", which depicts parts of the coasts of Carpentaria: based on Dutch sources, it gives 25 coastal names. In 1669 Richard Blome's map of the world was one of the first to attempt a representation of the whole of Australia. It is a copy of Sanson's map. William Berry, about 1680, likewise copied Sanson. But apart from these two publishers, English cartographers, in their first efforts in this field, resisted French theory for some time, basing their productions on the more correct early Dutch models, that is, on Tasman's Australia, without additions. This was done by Seller in 1675, Thornton in 1708, and by Senex, Moll, Grierson, Parker, Price and Willdey, and Overton and Hoole between 1711 and 1730. This representation was continued in the world maps of Mount and Page and the earlier world maps of Eman. Bowen (1747). Later, about 1760, Bowen succumbed to the French school, and Samuel Dunn in 1757 reproduced all the errors of Sanson.

Captain Cook

Cook's *Voyages* were published in 1773–84 and thereafter the east coast was correctly shown and the tracks of his ships were marked on many maps of the period; for example, in world maps of Sayer and Bennett (1772, 1781), Vaugondy (1773), Forster (1777), Richmond (1779), Zatta (1779 and 1787), Faden (1783, 1786 and 1790), Kitchin (1787), Bowles (1790), Wilkinson (1791), Laurie and Whittle (1794 and 1799), Cary (1801, 1811 and 1819), Walsh (1802), Thomson (1816), Walker (1818), Teesdale (1831) and Arrowsmith (1834).

The Naming of Australia

Quite an interesting study can be made of the naming of the continent. In its earliest-imagined form it was named "Australia Incognita," from the time of the Dutch Discovery it was called "New Holland," and this continued for some time after Cook's death. William Faden was one of the first to revert to the ancient title and name the whole continent Australia in 1810, and he was followed by Wilkinson in 1822 and Wyld in 1823, but its appellation remained fluid for some time, being still lettered New Holland by Thomson in 1824, Teesdale (1831) and Arrowsmith (as late as 1834) naming it New Holland or Australia. The Gemane Reilly lettered the continent Ulimaroa in 1795, as likewise did Cantzler in the same year and Reinecke in 1803 and Streit 1817.

Tasmania joined to Australia

Cook left a relatively small gap in the coast-line unsolved, the relation of Tasmania to Australia and the eastern part of the south coast. The early Dutch map-makers left Tasmania isolated, plotting the known coasts only on their maps. Vaugondy was one of the first to connect Tasmania with the mainland in his world map of 1752. This was also done by Bellin in 1753, and this continued up to the time of Bass's discovery in 1797, for example, by Vaugondy again in 1757, Janvier, Bowen and D'Anville (1760), Denis (1764), Bowles and Lotter

(1780), Faden (1787), Wilkinson (1791), Sayer (1792), Reilly (1795), Laurie and Whittle (1797).

The Hypothetical South Coast

In spite of the repeated failure of their hypotheses, cartographers in the main continued to give a conjectural outline, mostly by means of a dotted line, to the still unknown eastern portion of the south coast, among others Bowen (1777), Sayer and Bennett (1781 and 1787), Faden (1783, 1786 and 1790), Bew (1784), Kitchin (1787), Bowles (1790), Wilkinson (1791), Stockdale (1792), Laurie and Whittle (1794), Plant (1793), Reilly (1795), Cantzler (1795), Laurie and Whittle (1797), Richmond (1799), Cary (1801, 1806, 1811 and 1819).

A few to leave this part of the coast uncharted were Laurie and Whittle in 1799, Faden in 1802, Reichard in 1803, and Laurie and Whittle in 1808.

The Interior

No inland features were mapped from the time of the Dutch discovery in 1605 right up to the end of the 18th century when one or two coastal mountains were named. Rivers began to be indicated in the first decade of the 19th century, Thomson's map of 1814 being one of the earliest to indicate the general mountainous nature of the east and south coasts, and state boundaries began to be inserted after 1840.

Regional Maps

Maps of states or localities began to appear from about 1810 onwards; for example, Booth's Plan of the Settlements (1810), Oxley's chart of the Interior of New South Wales (1820), Mitchell's Map of New South Wales, Sydney (1834), Evans' Chart of Van Diemen's Land 1822). In 1832 J. Arrowsmith printed separate maps of the various states in his General Atlas of that date. This was also done by the Society for the Diffusion of Useful Knowledge in 1833 and by Tallis. Tallis's General atlas (1851) is one of the last of the decorative atlases, being adorned with small vignettes depicting the life and inhabitants of the region delineated. The first large map of Queensland on 4 sheets by T. Ham was published in Brisbane in 1856, Victoria on 2 sheets by Arrowsmith in 1853, Plan of the Town of Wellington by Captain Smith in 1840, and a large map of New Zealand by Wyld in 1840, and again in 1842, 1852 and 1861.

AUTHORITIES

CALVERT (A. F.). *The Discovery of Australia*, 4to, 1893. Second edition, 1902.

COLLINGRIDGE (G.). *The Discovery of Australia*, 4to, Sydney, 1895.

HARGREAVES (R. P.). *Maps of New Zealand in British Parliamentary Papers* (Otago Press, 1962).

—— *Maps in New Zealand Provincial Papers* (Otago Press, 1964).

HOCKEN (T. M.). "Some account of the earliest maps relating to New Zealand" (*Trans. and Proc. N.Z. Inst.*, 1894).

MALING (P. B.). *Early Charts of New Zealand*, Wellington, 1969.

Remarkable Maps of the XV, XVI and XVII Centuries (Amsterdam, Müller, 1894-7).

TOOLEY (R. V.). *Printed Maps of Tasmania* (Map Collectors' Circle, London, 1963).

—— *One Hundred Foreign Maps of Australia 1773-1887* (Map C. C., 1964).

—— *Early Maps Australia, Dutch Period* (Map C. C., 1965).

—— *Printed Maps of New South Wales* (Map C. C., 1968).

—— *Printed Maps of Australia*, 7 Parts 1970-73.

WIEDER (Dr. F. C.). *Monumenta Cartographica*, 1925-34.

—— *Tasman's Kaart van zijn australische ontdekkingen 1644 'De Bonapartekaart'* ('s Gravenhage, 1942).

WYND (I.) and WARD (J.). *A Map History of Australia* (Melbourne, 1963).

96 G. MERCATOR'S SCANDINAVIA ($18\frac{1}{2}'' \times 14''$)

from his *Atlas*, 1619

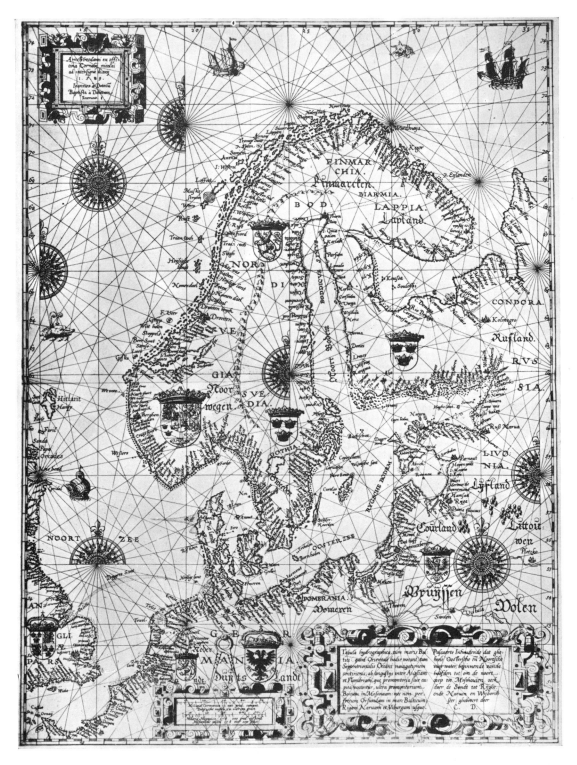

97   MAP OF SCANDINAVIA ($15\frac{3}{4}'' \times 21\frac{1}{4}''$)

from L. J. Waghenaer's *Den Nieuwen Spieghel der Zeevaert*, 1596

CHAPTER XIV

Scandinavia

A KNOWLEDGE of Scandinavia, as with other countries, was first revealed to the civilised world through Ptolemy (A.D. 150). As Ptolemy's limits were 83° N. he only gave indications for Jutland (Cimbric Chersonese) and the tip of the northern peninsula shown as an island (Skandia).

In the 14th century the first improvement in the mapping of these regions began, in the portolan charts of the Italians and Catalans (e.g. Dalorto's chart of 1330 gives a remarkable representation of Jutland for that date).

The earliest known map to deal with the Northern Regions was that compiled in manuscript in 1427 by Claudius Clavus (Claus Claussen Svart), a Dane resident for some time in Rome. An amended version, c. 1467, is found in another Ptolemy Codex now preserved in the Biblioteca Laurenziana in Florence. The Clavus map was the dominant map of the 15th and early years of the 16th century, for it was used, with some modification, in the 1482 Ulm edition of Ptolemy, re-issued 1486. Both these maps were from wood-blocks. It was again re-issued in 1507, this time in Rome, engraved on a copper-plate, and in subsequent editions up to 1535 (see Chapter I).

Another 15th-century map, one of Central Europe, including Denmark, Southern Scandinavia and the Baltic, was that compiled by Cardinal Nicolas Cusanus and printed posthumously in 1491. It survived in the wood-cut map in Schedel's *Liber Chronicarum* of 1493.

The first map of note in the 16th century was that composed by Jacob Ziegler, a Bavarian astronomer who while in Rome was on friendly terms with the archbishops of Trondheim and Upsala. Ziegler's map appeared in "*Schondia*," printed in 1532 (the first printed map to show and name Finland). Ziegler's map was used in part by Sebastian Münster for the fine map of Scandinavia in the various editions of his *Cosmographia* (see Chapter IV). Apart from the Ulm Ptolemys, which are exceedingly rare, Münster's map is the oldest wood-cut map obtainable of Scandinavia. Ziegler's map was also copied by Gastaldi in his small edition of Ptolemy 1548.

In 1539 a remarkable map appeared, the Carta Marina of Olaus Magnus of Upsala, printed in Venice on 9 leaves from wood-blocks. It is one of the notable maps of the 16th century, the first large-scale map of a European region, embellished with over 100 miniature engravings showing costumes, customs and fauna, embracing Norway, Sweden, Finland, Denmark and Iceland. Only one example of the original is known to survive, in Munich State Library. Fortunately the Carta Marina was re-issued, though on a reduced scale and from a copper-plate, by Lafreri, in Rome, in 1572 (seven surviving examples, only one in private hands). The map of Olaus was also used in his own *Historia*, printed in Basle in 1567 by S. B. Fickler, and largely utilised by later cartographers including Mercator.

A further advance was made in 1543 in the "Caerte van Oostlandt" of Cornelis Anthonisz. No copy of the original is now known, but a unique copy of the second issue of the map by Arnold Nicolai of Antwerp is preserved in Helmstadt. It is a wood-cut on 9 leaves. The work of Anthonisz was copied by Tramezini in Rome 1558, engraved by Jac. Bossius Belga, by Camocio in Venice 1562, and, with slight alterations, by Ortelius 1570, and Mercator 1595.

A cartographer of note was Marcus Jorden, a Holsteiner, professor of mathematics in Copenhagen. His map of Denmark of 1552 is now lost, and his map of Holstein of 1559 is only known from a single copy in Leyden. It was printed in Hamburg from a wood-block. The fine map of Denmark in Braun and Hogenberg's *Civitates Orbis Terrarum* (1585), the only map in that great 16th-century collection of town plans, was compiled by Jorden. In this same work (the *Civitates*) are three views of northern towns, Bergen, Stockholm and Copenhagen. All re-issued in 1657 by Jansson in his *Theatrum Urbium*, with Elsenor and Ripen, Freti Danici, Epis. Othenaras, and Visbia added. The influence of the Jorden map was considerable as it was adopted by Ortelius for the various editions of his *Theatrum*.

Another map is that by the Zeno brothers in 1558. This map of the north deceived their contemporaries and even enjoyed a spurious reputation in modern times, till the discovery of the original Olaus in 1886 and the Anthonisz in 1903 revealed the true source of their supposed knowledge. A more interesting production was the miniature Atlas *"Cosmographia"* of J. Honter, the first to give representation of separate Danish isles.

A fine map was produced by Lieven Algoet, "Terrarum Septentrionalium" (a copper-plate on 9 leaves) in 1562. Again only one example is known, in the Bib. Nationale, Paris. Algoet's map was used by De Jode in his Atlas of 1578 and again in 1593. Both editions are rare and the map is practically unobtainable.

No history of Northern map-making of this period would be complete without some mention of Tycho Brahe, the Dane. His actual cartographic output was small, but he stood head and shoulders above all his contemporaries as the greatest scientist of his age. His observatory on the island of Hven was the focal point for students and scientists of all nations from 1576–97. His methods and geographical teaching were spread through Europe by the practical application of his pupils, among others the great Blaeu. In the Atlas Major of the latter 13 plates are devoted to Hven and its instruments.

The year 1570 is for Scandinavia, as for other lands, a turning point in the history of cartography. In that year, on 22nd May, the *Theatrum Orbis Terrarum* of Abraham Ortelius was published, the first uniform world atlas. As maps in book form stand a far greater chance of survival than loose sheets, maps from this period onward, though still rare, have survived in greater quantities than those that preceded them.

Ortelius besides being the first in the field was one of the most enterprising of publishers, keeping his atlas up to date by adding new maps from time to time in succeeding editions. The *Theatrum* was printed in Antwerp by the great Plantin and issued both plain and coloured. The maps of Ortelius are always attractive, with large but fairly simple title cartouches and decorated mile scales, and usually painted in bright but semi-opaque colours. The maps of Scandinavian interest are as follows:

1570   Daniae regni typus. Septentrionalium regionem descrip.
1584   Daniae regni typus. Cornelis Antoniades descrip. Oldenburg comit. Laurentius Melchior descrip.
1590   Islandia (99). Ded. to Frederick II of Denmark. Andreas Velleius descrip. A. Ortel. excud. 1585.
1595   Cimbricae Chersonesi nunc Jutia descrip. Marco Jordano 1595.
         Holsatiae descrip. Marco Jordano Holsato auctore.

Of these the general map of the Northern Regions is the most attractive and justly popular. The first issue may be identified by the fact that the map is numbered 45 and the text on the back comes to a point with two words "Italice scripta." The other Latin editions of 1570 and 1571 are also numbered 45, but the text is carried straight across on the last line. The following points are a help in dating later editions of the map:

1571	Dutch	45	1581	French	81	1601	Latin	102	penult. line Sed Om
1572	German	45	1584	Latin	90				
1572	French	45	1587	French	90	1602	Spanish	105	
1573	Latin	60	1588	Spanish	90	1602	German	105	
1573	German	45	1592	Latin	97	1603	Latin	105	
1574	Latin	60	1595	Latin	102	penult. line Saxo	1606	English	102	
1575	Latin	60				1608	Italian	114	
1579	Latin	81	1598	German	71	1612	Latin	115	
1580	German	81	1598	French	103	1612	Italian	114	

The year 1584 witnessed an epoch-making event, the issue of Lucas Jansz Waghenaer's *Spieghel der Zeevaert*, the first printed Sea Atlas. Part II of the work issued in 1585 from the press of Plantin in Leyden included the northern navigation with the following charts all engraved by Joannes van Doeticum:

Sea coast of Norway between Noess and Mardou,
Mardou and Akersondt,
Distelberk and Waesberghe.
Map of the Sound.
Coast of part of Denmark and Sweden.
Sea coast of Sweden up to Stockholm.
Eastern half of the East Sea (including Finland).
West coast of Jutland.
Whole of Jutland.

In this first edition they appear with Dutch text on the back. Both parts were re-issued in 1585 with Dutch text, in 1586 with Latin text, in 1588 with English text (the plates re-engraved by De Bry, Hondius, Ryther and Rutlinger), again in 1588 with Dutch text, in 1590 with French text, 1591 Latin text, 1596 Dutch text, 1597 Dutch text, 1605 French text and 1622 English. The rarest edition is the first, with the French edition as rare. Then the English with black letter text, the most common being the Latin edition.

A new map, "Caerte van der Beldt," was added to the 1586 Latin edition and a new map of Norway, "Custe van Noorweghen (Bergen to Jedder)," in the 1588 Dutch edition. Two further maps of Scandinavia are to be found in the 1596 Dutch edition in the British Museum, both by Barentszoon, and are entitled "Tabula Hydrographica tam maris Baltici" (a fine map of the whole of Scandinavia) and "Hydrographica Septent. Norvegiae" (a map of the northern part of Norway) (97, 103). In spite of a fairly large number of editions Waghenaer is definitely a rare atlas. His charts are extremely decorative with large ornamental title-pieces, small figures of animals, ships and sea monsters. Waghenaer's charts are the first printed maps of separate parts of Scandinavia.

Two works on islands in the 16th century are of interest, the *"Isolario"* of Bordone and Porcacchi. The former has a map of Iceland, the latter a map of Iceland and one of Gotland. Towards the end of the century two miniature atlases appeared, one with maps of Ortelius which included a map of the Northern Regions, Livonia (showing coast of Finland), and Iceland. The Ortelius maps were issued by Peter Heyns with Dutch text, *"Spieghel der Werelt"* in 1577 and 1596, and with French text in 1583 and 1598.

A more interesting miniature with a longer life was the *"Caert Thresoor,"* first issued by

Barent Langenes in 1598 in Middelburgh and mostly engraved by Kaerius. It was re-issued in 1599 in Amsterdam by Cornelius Claesz and again with a fresh title in 1609. It contains the following maps of Scandinavia.

Denmark, Iceland, Northern Regions, Gotia (S. Sweden), Nortcaep (whole of Scandinavia), Livonia (showing Finland) and the island of Gothland.

A French translation appeared in 1602 and was re-issued in 1610. In these French editions (and only in the French edition) the map of the Northern Regions is replaced by a new map "Norbegia," showing Norway only. This is, as far as I know, the first printed separate map of Norway.

The *"Caert Thresoor"* continued its existence under a new name in the *Tabularum Geographicam* with Latin text, published by P. Bertius in 1600 and again in 1606. From the 1616 edition onwards (published by Hondius), however, new maps were given for the north, based on Mercator, viz. Polar Regions, Spitsbergen, Greenland, Iceland, Denmark, Sweden and Norway, and Gotia. Finally the original maps in Langenes were re-issued as late as 1649 by Visscher in his miniature atlas, and also with additions in an undated miniature by Blaeu.

To revert to the normal folio atlases, the last atlas of importance in the 16th century was the Atlas of Mercator, which appeared in its complete form in 1595. It contained the following maps of Scandinavia: Iceland, Sweden and Norway (96), Kingdom of Denmark, Northern Jutland, Funen, and Livonia (showing S. Finland).

Apart from the charts of Waghenaer and the miniature Honter these are the first large maps of separate parts of Denmark. These maps were used in subsequent editions of Mercator up to 1633 (see Chapter IV).

In Scandinavia as in England a stirring of cartographical interest took place in the 16th century, though in Scandinavia it remained in MS. form. For example, Rasmus Ludvigsson in Sweden was making rough plans for Gustavus Vasa during this period, and on the turn of the century Simon van Salingen, a Dutch merchant in the Danish Service, compiled an excellent map of the whole of Scandinavia in 1601.

The 17th century opens with a new and important map entitled "Nativus Sueciae adiacentumque regnorum typus," Jodocus Hondius Junior ex. Dedicated to Gustavus Adolphus by Adrian Veno (1613). This was the best map of Scandinavia up to its time, and set a standard followed for several years, as it was adopted with slight modifications in the Mercator–Hondius Atlas from 1632 onwards.

The most important atlas in the beginning of the 17th century was Blaeu's sea atlas. In spite of the number of editions and translations of this work it is, owing to its rarity, comparatively unknown, yet it is of the greatest interest for Scandinavia, for it contains no less than 12 maps of the northern coasts as follows:

> Der Noess to Oslo Fiord and south to Paternosters (101).
> Bergen to Der Noess. Dronten to Santwick.
> Noort Kyn to Dronten. Noort Cap to White Sea.
> Valsterbon to Upsala.
> Skagerack, Schagen to Copenhagen and Valsterbon to Maestrant.
> South Part of Belt. The Belt.
> Jutland, Rinkopen to Schagen. Jutland and Holstein.
> Coast of East Finland.

Blaeu's sea maps, published in oblong quarto, follow Waghenaer both in style and contents, and provide the link connecting the latter with the numerous marine atlases of the second half of the century.

In 1626 the first important separate map of Sweden appeared, by Anders Bure (Andreas Buraeus), with the title "Orbis Arctoi Nova et Accurata Delineatio." This was on a large

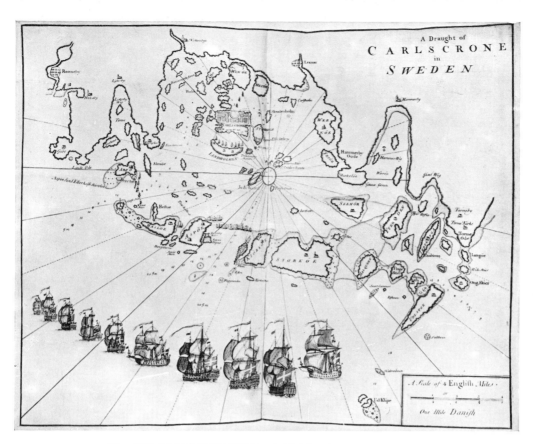

98   SIR JOHN NORRIS'S CARLSCRONE (21½″ × 17″)
from *Compleat Sett of New Charts*, 1723

99   ABRAHAM ORTELIUS'S ICELAND (19¼″ × 13¼″)
from his *Theatrum Orbis Terrarum*, 1590

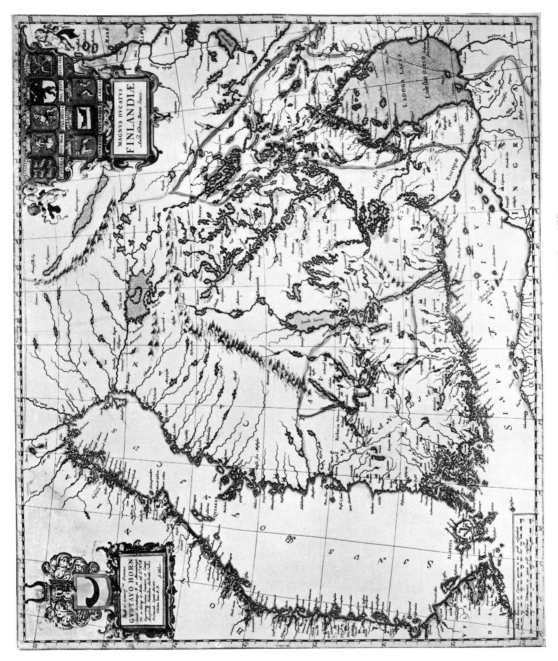

100 G. BLAEU'S FINLAND (20½" × 17")
from his *Atlas Major*, 1659

[129

scale, being printed on 6 sheets in Stockholm. Anders Bure had, however, compiled a map of the northern part of the kingdom even earlier, viz. in 1611 (only two copies known). Of the map of 1626 seven copies are known, all in Scandinavia, in various states as it was revised from time to time.

Bure's map was a thorough survey far in advance of any preceding effort, and his work was soon incorporated in contemporary atlases of Dutch and other publishers. Visscher, for example, evidently used Bure's map combined with other sources to produce a new type for Scandinavia. This was engraved by Abraham Goos in 1630, was adopted in the Mercator Atlas of 1633, and remained the standard map of these regions for many years, being used by Blaeu, Tavernier, De Wit, and many others. Bure's map was also used for regional work, Blaeu producing Upland, Gothia and Livonia in his atlas of 1641. He greatly increased the number of Scandinavian maps in his *Atlas Major* of 1662.

An even larger survey than Bure's was accomplished by John Mejer, a Holsteiner, who was Geographer Royal to the King of Denmark, from whom incidentally, owing to troublous times, he received scant encouragement. He had completed 37 maps of Schleswig-Holstein before 1648, and these were published by Danckwerth in 1652.

These numerous surveys led to a large output of printed maps in the middle and later part of the century. A brief résumé is as follows (for various editions, see Chapter IV, Holland).

GENERAL MAPS OF SCANDINAVIA: Mercator-Hondius, Jansson, Blaeu, Visscher, Sanson, De Jonge, Pitt, De Wit (1660) (another 1680), Danckerts, Lea, Berry, Jaillot, Merian, Wells.

NORWAY: General maps of the Kingdom. Blaeu, dedicated to King Christian, Jansson, Sanson (1668), Pitt (1680), De Wit (two states, one with one coat of arms only, Norway, and one with two coats of arms, Norway and Sweden), Danckerts, Valk and Schenk.

Stavanger. Blaeu, Jansson, Pitt.
—— (South). Blaeu (1662).
—— (North). Blaeu (1662).
Bergen Diocese. Blaeu (1662), Sanson (1668), Pitt (1680).
Trondheim. Blaeu (1662), Pitt.
—— (North). Sanson (1668).
—— (South). Sanson (1668), Pitt.
Finmark. Blaeu (1662).
Aggerhus. Sanson (1668).
Wardhus. Sanson (1668).

SWEDEN: General maps of the Kingdom. Blaeu, dedicated to Carolus Gustavus, Sanson (1654 and 1669), Pitt (1680), De Wit.

Scania. Blaeu (1662), Jansson, dedicated to Othoni Kragh, Johann Husman (1677), De Wit, Danckerts, Pitt.
Gothia. Blaeu, Jansson (1642), Pitt, De Wit.
  S. Gothlande. Sanson (1659).
  Westro Gothlande. Sanson (1666).
  Ostro Gothlande. Sanson (1666).
Sueonia. Blaeu, dedicated to Axelio Oxenstierna (1662), Sanson (1666), Pitt.
Noorland. Blaeu, dedicated to G. Bengtson Oxenstirna (1662).
Bahus. Sanson (1668).
Helsinge–Angermanni. Sanson (1666).
Lapland. Blaeu (1662).
  Lapland (East). Sanson (1666).
  Lapland (West). Sanson (1666).
Bothnia Orient. Sanson.

FINLAND: General maps of the Kingdom. Blaeu, dedicated to Gustavus Horn (100), Sanson (1666), Moses Pitt (1680), De Wit (1680), Valk and Schenk (1700).

Carelia Ingria and Ingermanland. Sanson (1666).
Savolax and Kexholm. Sanson (1669).
Cajanie or Bothnie Orient (1666).
Livonia (showing south coast of Finland). Ortelius, Mercator, Blaeu, Jansson, Pitt.

DENMARK: General maps of the Kingdom. Ortelius, Mercator, Speed (first edition 1627, printed Humble; later edition 1676, printed Basset and Chiswell), Hondius, Blaeu, Jansson (dated 1629, with border of costume figures and town views; another map on larger scale, without figured border, dedicated to Gerardo Schaep), Visscher, Dahlberg (1646), Sanson (1658), De Wit (1659), Moses Pitt, De Wit (1680), Danckerts, Berry (1683), Seller (1690), Jaillot (1692), Merian, De Fer (1700).

Jutland. General Map. Mercator, Blaeu, Jansson, Sanson, De Wit, Visscher, Pitt.
—— (Ripe to Aarhus). Blaeu (1662), Pitt (1680).
—— (Alburg–Viburg). Blaeu (1662).
Funen. Mercator, Blaeu, Jansson, Pitt.
Laland and Falster. Jansson, Blaeu, Pitt.
Zeeland. Jansson, dedicated to Georgius Seefeldus, Blaeu, dedicated to Petrus Charisius, Pitt.
Danish Isles. De Wit, Danckerts.
Schleswig–Holstein. Ortelius, Mercator, Hondius, Blaeu, Mejer (1650), Jansson, Pitt, De Wit, Visscher.
Bornholm. Atlas Danicus (1677).
Hven. Blaeu, dedicated to Christianus Longomontanus.

ICELAND: Ortelius, Mercator, Hondius, Blaeu, Jansson, Sanson, Pitt.

All the above maps are highly decorative, and all the Dutch in the same style. The title is enclosed within a large cartouche, usually with figures depicting the costume or industry of the country or region, heraldic coats of arms and decorated scales of miles with cherubs, etc., and a few small ships scattered in the sea. The French maps by Sanson are quite different, slightly smaller in size, their ornament being restricted to the title-piece (usually a martial design) apart from an occasional small ship. Their distinguishing features are the clarity of their engraving, the absence of overloading detail, and skill in lettering, making them the easiest of maps to read. The Englishman Pitt, working with his Dutch partners, is English only in name, his plates being Dutch plates. Blaeu's maps were undoubtedly the finest land maps of the 17th century, and are closely followed by Jansson. The above list can be greatly expanded, of course, as there were many issues of Blaeu's and Jansson's Atlases with different texts (see Chapter IV).

We now turn to an even more prolific and in some ways a more fascinating branch, the sea charts. Maritime survey invariably preceded and was far in advance of land survey on Scandinavia. Whereas land maps once made tended to remain static till a new type appeared, sea charts were constantly revised, either in small particulars or large areas, a fact that makes their study most absorbing.

Italy made a fine contribution in the magnificent *Arcano del Mare* of Sir Robert Dudley (1646), re-issued 1661; from Holland came atlases by Colom, Doncker, Goos, Jacobsz, Keulen, Van Loon, Robijn and De Wit (see Chapter V); from England Seller's *English Pilot* and Thornton's *Atlas Maritimus*, and from France the *Neptune Français*. All these contain charts of great interest for Scandinavia. From the "*Arcano del Mare*" come:

101   G. BLAEU'S CHART OF NORWAY ($21\frac{1}{2}'' \times 19\frac{3}{4}''$)
from *Le Flambeau de la Navigation*, 1625

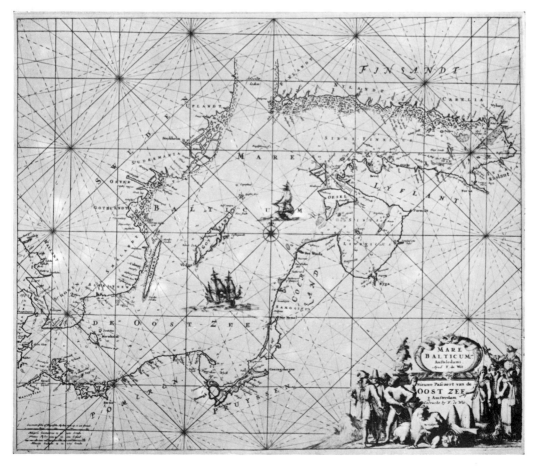

102   F. DE WIT'S BALTIC ($22\frac{1}{2}'' \times 19''$)
from his *Sea Atlas*, 1680

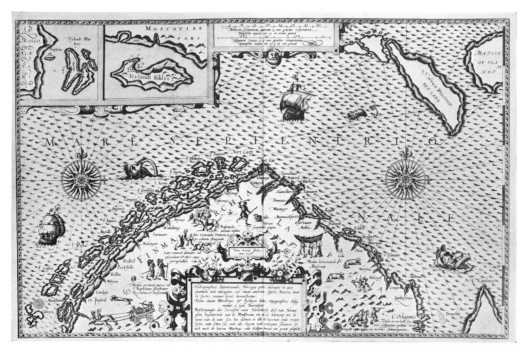

103   G. BARENT'S NORTHERN NORWAY (19¾″ × 12⅞″)
from L. J. Waghenaer's *Den Nieuwen Spieghel der Zeevaert*, 1596

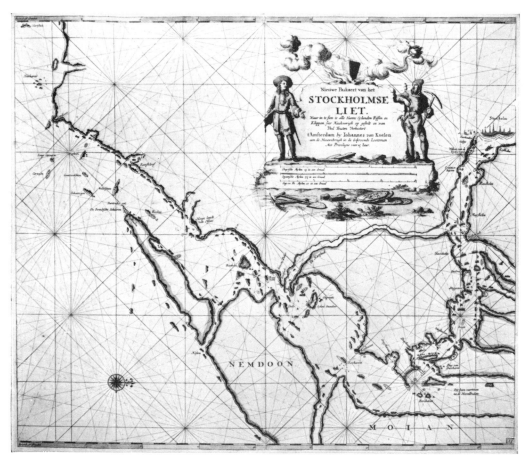

104   J. VAN KEULEN'S STOCKHOLM (23″ × 20″)
from his *Sea Atlas*, 1683

Carta di Norwegia piu moderna.
Carta particolare della parte Australe della Norwegia.
L'Isolle di Fero . . . con le Noruegia Settent. (Bergen–Lofoten.)
Carta particolare della costa di Finlandia (misprint for Finmark).
Carta particolare . . . Jutlandia . . . parte della Suetzia e della Noruegia.
Carta particolare dell Mare Baltico. (Eleeholm–Padus.)
Carta particolare del Mare Baltico. (Colbergen–Westueson.)
Carta particolare dell fine del Mare Baltico. (Abo–Wyburgh.)
Carta particolare della entrata del mare Botnico. (W. Finland and Sweden.)
Carta particolare . . . Albis . . . al Sondo di Danemarca.
Carta nonna generali di Europa. (N. Norway–White Sea.)

It is not possible in one short chapter to enumerate the differences between the various Dutch publishers and their English followers, but it is possible roughly to tabulate them, at the same time pointing out that few of them were slavish copies, almost all adding some new information or correction to some part or other of the map, either of outline or addition of new names, or both.

NORWAY:

Der Noess to Paternosters
    Doncker, Jacobsz, Robijn.
Der Noess to Schuitenes and Liet van Bergen
    Jacobsz, Robijn, Mount and Page, Seller.
Bergen to Hoek van Horrel
    Doncker, Jacobsz, Seller, Robijn, Mount and Page.
Hoeck van Horrel to Momendael with Liet Dronten.
    Doncker, Jacobsz, Robijn, Mount and Page.
Sanien to Noord Kyn
    Doncker, Jacobsz, Robijn, Mount and Page.
Noord Kyn to Kola River
    Doncker, Jacobsz, Robijn, Mount and Page.
Finmark. Pascaart vande zee custen van Finmarken, Laplant, Russlant, Nova Zembla en Spitsbergen
    Jansson, Doncker (2 editions), Goos, Van Loon, Jacobsz, Seller, Robijn.
Finmarchiae et Laplandiae Maritimae
    De Wit, Renard, Ottens.
Noort Zee
    Jansson, Doncker (3 different issues), Van Loon, Jacobsz, Robijn, Seller, De Wit, Mount and Page, Thornton, Renard, Ottens.
Noorwegen (Elsburg to Dronten)
    De Wit, Renard, Ottens.

SWEDEN and FINLAND:

Die Custen van Denemarcken en Sweden (Valsterbon–Calmer)
    Doncker, Jacobsz, Van Loon, Robijn.
Caarte van Sweeden (Oeland–Stockholm)
    Jacobsz, Van Loon, Robijn.
Stockolmse Liet and Gat van Abbo
    Jacobsz, Robijn.
Pas Caart van Liiflandt en Oost Finlandt
    Doncker, Jacobsz (2 charts).
Oost Zee
    Jansson, Doncker (1664) and (1665), Goos, Van Loon, Jacobsz, Seller, De Wit (102), Thornton, Renard, Ottens.

DENMARK:

Paskaert van't Schager–Rak
   Doncker, Jacobsz, Loots.
Oost Belt
   Doncker, Jacobsz, Seller.
West and Oost ʒyde van Jutlandt als mede de Belt
   Doncker.

Renard and Ottens, though both of the 18th century, have been included here as they continue the series. It can also be considerably enlarged by various editions of Mount and Page.

The most prolific chart makers of the end of the 17th century and early 18th century were, however, the firm of Keulen. Composed in the traditional Dutch style with large and charming vignette title-pieces showing the costume and occupations of the area delineated or local fauna, decorated scale of miles with Neptunes, Tritons, etc., occasional small ships, whales, compass roses, etc., Keulen's charts relating to Scandinavia are as follows:

Noord Zee (title in Dutch, English and French) (shows W. Denmark, Coast of Norway, E. coast of England).
—— (Calais to Landt van Stadt, Norway and Kent to Fero Is.).
—— Der Neus–Nova Zemla (including Iceland, Spitsbergen and North Scotland).
—— Tessel na de Zuyd Kust van Noorwegen (Capt. J. Heytemann).
Noordlyckste Deel van Europa (Scandinavia, Nova Zembla, Spitsbergen, Iceland, Greenland and inset Jan Mayen Is., $23 \times 20\frac{1}{2}$ ins.).
Noorder deel van Europa . . . Groenlandia en Moscovise Scheepvaard.
Noorder deel van Europa (Finmark, Lapland and Spitsbergen, 3 insets, $39\frac{1}{2} \times 23$ ins.).
Noorder deel van de Noord Zee. 1726.
Noord Occiaen, Hitland–Straet Davids (Iceland, Greenland and coast of America).
——, Terra Nova–Straat Davids (including Iceland).
Iceland, Spitsbergen and Jan Mayen Is. together on one sheet (1683), re-issued with additions (1716).
West Kust van Jutland, Busem–Jutsche Riff (1683), re-issued with additions (1716).
Jutlandt, Jutsche Riff–Schagen (with part of Norway, Der Neus–Oxefoort). (1683), re-issued with additions (1716).
Noorwegen, Oxefoort–Gottenborg. (1683), re-issued with additions, dated 1716.
—— Der Neus–Bommel Sond, and t'Liet van Bergen.
—— Bommel Sond–Wtwer Klippen, and t'Liet van Bergen.
—— Wtweer Klippen–Swartenos.
—— Swartenos–Heyligelander Leen, and Liet van Dronten.
Finmarcken, Heyligelander Leen–Tromsondt.
—— Tromsondt–Tiepener.
Kust van Noorweegen, Lansgesond to Gottenborg (8 insets of harbours, 36×20 ins.).
[Het inkoomen en Reeden van Gotten Borg] (36×20 ins.).
Schager Rack, Schagen–Copenhagen and Gottenborg–Malmuyen (with 8 plans of harbours, $34\frac{1}{2} \times 20$ ins.).
—— Schagen–Elseneur and Pater Nosters–Elsenborg.
—— Schagen–Copenhagen and Maardou–Landscroon (dedicated to L. Erasmus Laasbye, 39×23 ins.).
—— door C. J. Vooght.
—— de Sond ende Beld . . . door Nicolaas de Vries.
Sond ende Beldt (Zeeland, Fuynen, Lalandt, etc.) (1683), re-issued with additions (1716).
—— door C. J. Vooght. .
De Sond int Groot (East Zeeland and Sweden from Engelholm to Falsterbon).
Oost Zee, Rostock–Wyborg (General Chart of the Baltic, parts of Sweden, Finland, Coerland, etc.).
—— Valsterbon–Schenkkenes (1683), re-issued with additions (1716).
—— Schenkkenes–Stokholm (1683), re-issued with additions (1716).

Stockholmse Liet (104).

Oost Zee, Sernevisse–Parnout (shows part of Gotlandt (1716).

—— Lemsaal–Beooste Kok (and coast of Finland, Nykerk–Borgo, 1683, and re-issued with additions 1716).

—— Broklom–Strellen (and coast of Finland, Parna–Wyborgh, 1683, and re-issued with additions 1716).

—— ... nieu opgestelt door Nicolaas de Vries (inset Ladoga, $23\frac{1}{2} \times 20\frac{1}{2}$ ins.).

Nieuwe paskaert voor een Gedelte van de Oost Zee (39×23 ins.).

[Oost Zee oster Sioon] (Ingelstad–Christianopel, 31×20 ins.).

Nieuwe afteekening van de Finlandse Golf of Boden.

Several native cartographers were at work during the latter part of this period, e.g. Johann Mansson, Ornehufud, Olasson Feterus, Erik Dahlberg, Tresk, Petter Gedda, Gripenhelm, etc.

The 18th century was not so spectacular as the 17th. For one thing atlases became smaller in volume, if not in size, and so Scandinavia was limited usually to a general map of the whole kingdom, and maps of the provinces disappeared. There were two exceptions however, Homann in Nürnberg and Seutter in Vienna, who both issued large atlases in the previous style, producing maps of Denmark, Jutland, Danish Isles, Sweden, Norway, Bahus, Carelia, etc., and views of Stockholm and Copenhagen. In the beginning of the century large maps of the whole peninsula were popular, e.g. Jaillot in France, Bowen, Moll, Senex, Berry and Grierson in England. The most popular of these is Moll's Denmark and Sweden, with its broad side borders depicting costumes and customs of the Laplanders. There were also some large-scale maps of merit, e.g. Capt. Hammond's North Sea, 2 sheets (1720), Moll's Northern Navigation, 4 sheets, Wilson's Baltick, 4 sheets (1720), in France Bellin's large charts, and in Holland De Leth's Gulf of Finland. Of small maps of the four kingdoms, the most attractive are those at the beginning of the century by Peter van der Aa, and right at the other end of the century, by A. Zatta. The English maps of Faden are likewise finely engraved. There were numerous English Pilots of this time (see Chapter VII), and one work deserving special notice is Sir John Norris's *Sett of New Charts* (first edition 1723) (98), 20 maps devoted exclusively to Scandinavia. Finally Kongl. Landmaterj Contoiret, Pontoppidan (1781–5), Hermelin, and Wessel and Skanke, at the end of the century, were amongst the last to produce maps that were decorative as well as useful.

AUTHORITIES

AHLENIUS (K.). *Till Kannedomen om Skandinaviens Geografi och Kartografi under 1500 talets senare halft* (Upsala, 1900).

BAGROW (L.). *A. Ortelii Catalogus Cartographorum*, 2 vols., Gotha, 1928–30.

BJORKBOM (C.). *Robert Dudley's Sjöatlas* (Segel och motor, 1937).

BJORNBO (A. A.) and CARL S. PETERSEN. *Anecdota Cartographica Septentrionalia* (Hauniae, 1908).

—— —— *Cartographia Groenlandica 1000-1576* (Kobenhavn, 1912).

BLAU (J.). "Mémoire sur deux monuments géographiques conservés à la bibliothèque publique de Nancy" (*Mém. de la Soc. Roy. de Nancy*, 1835).

COLLIJN (I.). *Olaus Magnus Gothas ain Kurze auslegung der hennen Mappen von den alten Goettenreich und andern Nordlenden* (Stockholm, 1912).

DAHLGREN (E. W.). "Gamla tyska kartor i Kungl. Bib. i Stockholm" (*Nordisk. Tids. f. Bok och Bibl.*, 1914).

DAHLGREN (P. J.) and H. RICHTER. *Sveriges Sjökarta* (Stockholm, 1944).

DE LA GARDIE (M. G.). *Samling af aldre stadsoyer och Historiska Planscher i Kungl. Biblioteker. Forteckning i Collijn* (Stockholm, 1915).

EGGERS (C. U. D.). *Physikalische und statistische Beschreibung von Island* (Kopenhagen, 1786).

ERSLER (E.). *Jylland. Studies of Skildringer til Danmarks Geografi* (Kiobenhavn, 1886).

FAGGOT (J.). *Historien om Svenska Landtmateriet ock Geographie* (Stockholm, 1747).

FELLMAN (A.). *Voyage en Orient du Roi Erik Ejegod et sa mort à Phapos* (Helsinki, 1938).

GORDIN (I.). *Old Maps of Finland* (Helsinki, 1967).

HERRMANN (H.). "Die Länder des Nordens in Kartenbilde vom Altertum bis zum 19 Jhr." (*Der Norden* 16, 1939).

HERMANNSSON (H.). "Two cartographers Gudbrandur Thorlaksson and Thordur Thorlaksson" (*Islandica*, Vol. XVII, Ithaca, N.Y., 1926).

—— *The cartography of Iceland* (Islandica, Vol. XXI, Ithaca, N.Y., 1931).

JOSEPHSON (R.). *Olaus Magnus om den nordiska Konsten* (Kulturminsher, 1940–1).

KNUDSEN (J.). *Het Leeskaartboek van Wisbug . . . inleidung door Dr. C. P. Burger* (Kobenhavn, 1920).

KOHLIN (H.). "Georg von Schwengeln and his Work, 1620–1645" (*Imago Mundi VI*, 1949).

KEUNING (J.). "Cornelis̆ Anthonisz" (*Imago Mundi VII*, 1950).

LAURIDSEN (P.). "Kartografen Johannis Mejer" (*Historisk Tidsskrift*, Kobenhavn, 1888).

LONBORG (Sven). *Sveriges Karta tiden till omkring 1850* (Uppsala, 1908).

LYNAM (E. W. O'F.). "The early maps of Scandinavia" (*Geogr. Jnl.*, 1927).

—— *Early Maps of Scandinavia and Iceland.* Reprinted from Saga Book of Viking Society for Northern Research, 1934.

—— *The Carta Marina of Olaus Magnus Venice 1539 and Rome 1572* (Jenkintown, 1949).

NORDENSKIÖLD (A. E.). *Facsimile Atlas* (Stockholm, 1889).

—— *Om broderna Zenos resor och de aldsta kartor ofver Norden* (Stockholm, 1883).

NORLUND (N. E.). *Danmarks Kortlaegning en historisk Fremstilling* (Kobenhavn, 1943).

RICHTER (H.) and W. NORLIND. *Orbis Arctoi nova et accurata delineatio auctore Andrea Bureo Sueco 1626, text and atlas* (Lund, 1936).

—— and H. KOHLIN. *Gamla Karto varlden Norden Vustkusten* (Goteborg, 1948).

SOC. DE GÉOG. DE FINLANDE. *Travaux Géogr. exécutés en Finlande* (Helsingfors, 1895).

SPEKKE (A.). "A Brief Cartographic/cosmographic view of the Eastern Baltic Coast up to the 16th century" (*Imago Mundi V*, 1948).

THOMASSY (R.). "De Guillaume Fillastre considéré comme géographe" (*Bull. de la Soc. de Géogr.*, 17, 1842).

WAITZ (G.). "Des Claudius Clavius Beschreibung des Skandinavischen Nordens" (*Nordalbingische Studien*, Kiel, 1844).

WALLIN (Vaino). *Suomen Maantiet. Ruotsin vallam aikana* (Kuopio, 1893).

Index